I0531190

Old Man
of the Fossil Beds

Praise for the book

OLD MAN OF THE *Fossil Beds* is a remarkable book by a remarkable author.

Melanie Bonner Thomas grew up in western Kansas, the youngest of eight children in a renowned and accomplished family of amateur paleontologists whose findings are housed in natural history museums across North America.

For those not familiar, that part of Kansas sits atop remnants of the North American Inland Sea, a prehistoric continent-splitting body of water that left deep deposits of fossil-rich sediments in chalk and limestones beneath the Great Plains.

To this land of dry wind and buried treasures comes Marion "Skeet" Bonner, a reedy, young boy with boundless energy and an endless fascination with the gargantuan fish and toothed birds that existed during the Late Cretaceous, some sixty-seven million years ago.

As tough and gritty as the land itself, Marion Bonner is, by turns, scholarly, stubborn, charming, funny, industrious, alcoholic, dominating, gambling-prone, musical, and about as beloved as a father could possibly be.

Of scientific matters the author writes in economic, accessible prose with confident precision. She makes us feel the excitement and importance of the past worlds that lie beneath our feet. But it is as a memoirist that Bonner Thomas really shines. She was nine years old when her mother died, leaving the young girl to come of age under the tutelage of her eccentric father and attentive older siblings. Portraits of the family, the Western Kansas community, and academic notables who made pilgrimage to the Bonner household, are honest, touching, and beautifully drawn.

Trust me, it's rare to find a writer who demonstrates equal facility with technical details and the flesh and blood of humanity. Melanie Bonner Thomas succeeds admirably in both areas. Her readers will find themselves both educated and moved by the *Old Man of the Fossil Beds*.

—Charles Forrest Jones, author of *The Illusion of Simple*, former director of the Kansas University Public Management Center

Old Man
of the Fossil Beds

**A Kansas Dreamer, His Family of Hunters,
and Their Search for Prehistoric Sea Creatures**

Melanie Bonner Thomas

chalk lily books

Copyright © 2025 by Melanie Bonner Thomas

All rights reserved.

This book is nonfiction. It is designed to provide accurate and authoritative information in regard to the subject matter covered. So far as possible, statements of fact are supported by original source material including books, newspapers, magazine and scholarly articles, personal papers and correspondence, and field notes. While the publisher and author have used their best efforts in preparing this book, they make no representations or warranties with respect to the accuracy or completeness of the contents of this book.

No part of this book may be reproduced in any form without written permission from the publisher or author, except as permitted by U.S. copyright law.

Identifiers:

ISBN 979-8-9930531-2-7 (hardback)
ISBN 979-8-9930531-0-3 (paperback)
ISBN 979-8-9930531-1-0 (ebook)

Library of Congress Cataloging-in-Publication Data
LCCN 2025919484

Book cover and Chalk Lily logo designs by Donna Neal, *www.firethornstudio.com*
Cover photo of Marion Bonner in the *Bonnerichthys* quarry by Charlie Norton.

Chalk Lily Books, LLC
P.O. Box 13145
New Orleans. LA 70185-3145
www.chalklilybooks.com

chalk lily books

Dedication

For my brother Ralph Stephen Bonner, who joined Dad
in fossil hunter heaven on September 27, 2023.
I am grateful for Steve's insights and suggestions
during the drafting of this book in 2023.

He leaves a void in all of us and is deeply missed.

Photo by Ashley Bonner Zung, 2018.

Credits

Most of the photographs included in this book are from the Bonner family albums, taken by Bonner and Berg ancestors, primarily my mother, Margaret (Berg) Bonner. Many of the later ones were taken by my siblings and credited when I could determine the photographer. Noncredited recent photos are by the author. Except for cropping, photos have not been digitally altered.

The representations of extinct Cretaceous Period sea creatures in the spread on pages 218 and 219 at the end of the book were painted by my brother Chuck Bonner as a mural located in the Keystone Gallery and Fossil Museum in western Kansas.

My father saved all correspondence and books he deemed important. He didn't have a home phone until 1965, so he made most of his early arrangements with museums by mail. This book could not have been prepared without reams of Bonner family primary sources: letters, cards, notes, drawings, newspaper clippings, and even packing slips for fossils. Without these documents, I would not have been able to capture details of my father's and my family's story, memories notwithstanding.

A note on the book's approach: Because this work combines scientific biography with family memoir, each of its four parts begins with a personal narrative chapter. Therefore, the presentation is not chronological. To help the reader, each chapter's subtitle indicates the decades in which most of the action occurs.

There rolls the deep where grew the tree.
O earth, what changes hast thou seen!
There where the long street roars, hath been
The stillness of the central sea.

The hills are shadows, and they flow
From form to form, and nothing stands;
They melt like mist, the solid lands,
Like clouds they shape themselves and go.

But in my spirit will I dwell,
And dream my dream, and hold it true;
For tho' my lips may breathe adieu,
I cannot think the thing farewell.

— Alfred, Lord Tennyson, *In Memoriam,* canto 123

Little Jerusalem Badlands State Park, Logan County. The chalk has been eroded by wind and water to become bluffs and canyons. The Bonner family hunted this site decades before it was declared a protected area. Photo by Carlton Thomas.

Contents

PART I - BEDROCK

Previous page: The Bonner fish-within-a-fish, a complete *Gillicus* inside a complete *Xiphactinus*, as it appears on display in the Royal Tyrrell Museum, Ontario, Canada. Marion's youngest son, Dana Bonner, discovered this specimen in 1982. Photo courtesy of the Royal Tyrrell Museum.

Plains Vision
1980s

HE LOOKED AT THE prairie and saw the sea. For my Dad, the past was the present. The sea was the Niobrara Cretaceous inland sea that once flowed over Kansas.

Anyone still alive who has heard about Marion Charles Bonner, a self-taught paleontologist from western Kansas, remembers his claim to fame was fossils. Anyone who knew him as a friend, father, or grandfather remembers the fossils but so much more: poetry, songs, funny stories told with a corny or earthy sense of humor. He was a man of the plains, a performer, and a lifelong student of history and science.

I remember him as a nontraditional, nurturing father whom I often called "Pop" and "Poppa." My parents, Marion and Margaret Bonner, had eight children from the years 1936 through 1957. I was the youngest, and Pop raised me alone after our Mom died when I was nine. My four brothers, three sisters, and I all grew up in the small western Kansas town of Leoti, pronounced "Lee-*oh*-tah," in Wichita County, a sparsely populated county in a sparsely populated state. Both county and state are named after Native American tribes. The number of people in the entire state of Kansas is not much more than that of greater Chicago.[1]

Following our graduation from high school, we Bonner children scattered from Wichita County like so many tumbleweeds, but our father, a widower from 1967 on, stayed put. As spread out as his children were, it never occurred to him to leave western Kansas. The place was his identity, and as long as he was able to, he would climb into his car and roll out to the fossil beds. His personality was textbook Kansan, exemplifying the solidness described by Kansas newspaper publisher and U.S. Senator Arthur Capper: "Kansans 'stay on their feet more firmly, keep their heads, think more sanely than any people I know.' "[2] But as resolute as Pop was, he was far from being a typical Kansan. He also joked about everything, including his state, frequently repeating a comic song about Kansas that he had memorized as a teen:

The sun's so hot that eggs will hatch
Way out west in Kansas.
It popped the corn in a popcorn patch
Way out west in Kansas.

An old mule coming down the stretch
Saw the corn and caught his breath
He thought 'twas snow and froze to death
Way out west in Kansas.[3]

During my college years, I often drove "way out west" to visit Dad, gathering tidbits to use in my master's thesis at the University of Kansas in Lawrence, on the east side of the state. I shared with him the research I was doing on a subject he had kindled in me as a child. My thesis was an annotated version of W. E. Webb's 1872 book about western Kansas called *Buffalo Land,* which Dad and Mom had read in the 1960s.[4] Pop had moved from Leoti and was living in his final home in Healy, in Lane County. I saw him as a resource and liked to chat with him about Kansas history.

One day in April of 1982, I drove out to Healy, located seventeen miles north and west of the Lane County seat of Dighton, for a weekend visit. This time, instead of sitting at the dining room table discussing western Kansas history, I asked to go to the fossil beds. Dad was an old hand at hosting visiting friends, scientists, and writers who came to his house to ask about his fossil hunting experiences. He was more than happy to fix a picnic lunch and take me out touring, driving from Lane to Logan County on that warm, clear Saturday.

For those who live in any town or urban area, coming out to the semi-arid plains of western Kansas requires a few hours of adjustment. I always felt somnambulant at first. The smell of dry grass, the hum of tires on warm pavement, Dad's voice droning over oft-retold tales, had the effect of lulling me into a stupor. At first, the plain itself seemed monotonous and empty. But as geologists and paleontologists know, the plains of western Kansas are not boring. There are millions of stories lying in the sediments under the surface.

As Pop and I made our way to the fossil beds with the windows of his Oldsmobile Delta 88 down, the warm, dry air filtered into my lungs. In the distance, heat waves rose from the road ahead. The wind ruffled Dad's long white hair behind his seed cap, revealing a neck that was wrinkled, tan, leathery. Gazing out the window as we sped through the vast landscape, my blue bandanna whipping in the air, I could see for miles in every direction. The higher ground was covered with grama and soapweed. Fences on each side of the road were jammed with gray Russian thistles, a.k.a. tumbleweeds.

"Look over there!" Pop said excitedly, his eyes focused on grass-covered hills two hundred yards away. Weathered fence posts strung with barbwire were what I saw first; then I looked farther afield and saw the animals—seven pronghorns. They stood frozen in place for a moment, then turned and bounded away, white rumps easily gliding over another fence on the far side of the pasture. Their presence caused me to waken to the stark beauty of the land, snapping me out of the young person's boredom with the old person's stories.

Something about the place's desolation and Dad's simple statements about local history during the ride made me imagine we were moving across the plains one hundred years in the past. He was the stagecoach driver and I was riding shotgun, bumping along. "See that coyote over there?" he asked. "Where?" I clamored, "I don't see a thing!" Dad pointed with his middle and partial index finger at the horizon. "Look! See that black dot? That's a coyote." Indeed, as I focused, I saw a small dot moving westward.

Bronze bust of Marion Bonner created in 1989 by western artist Charlie Norton of Leoti, Kansas.

Approaching the pasture that contained the outcroppings he called Hell's Half Acre, he slowed the car down, gravel road crunching. The destination was a corner entrance to the pasture. We would go in through a barbwire gate, adhering to one of the Bonner "fossiling" rituals. The passenger unhooked the gate and walked it open, Dad drove through, then the passenger closed the wire gate behind the vehicle. Since one of my brothers typically did the gate handling, Dad got out and checked to see if I had properly placed the weathered pole in the loops and secured it carefully. He pulled himself to his full height, slightly shorter than his youngest daughter's height of five feet eleven. "You're checking up on me?" I asked, teasingly. "Honey, I know you can do it, but sometimes these loops are a little tight." He tugged at it, gave me a blue-eyed glance, then a thumbs up, and said, "That's a dandy job! Someone taught you how to do it right." We both chuckled and got back in the car.

Dad navigated slowly over the clumpy grass toward the area where chalk bluffs were exposed to the elements. Even though he carefully avoided the rutted cow paths, bush-high prickly pears, and large yucca, it was rough going. Each jolt of the car shook us, scrub brush scraping the car's undercarriage. The light green Olds blended nicely with the spring grasses. As we approached some rangy Hereford cattle munching lazily on grama grass, a few of them turned and rambled away from the slow car. "Hell, it's a stampede!" he joked.

We parked on a rise overlooking the chalk washout. We both grabbed a canteen and Dad opened the trunk to perform the ritual handing out of fossil tools. He gave me a nice scraper pick, and I located a favorite brush. Dad and I spent the morning roaming the dry beds, scanning for fossils. Walking along slowly, he sang an old Scottish ballad: "You take the high road and I'll take the low road, and I'll be in Scotland a-fore you."[5] This was his way of signaling that I should focus my hunting on the upper part of the bluff, while he strolled along the lower part of the bluff, checking for bones that might have washed down into the gully.

We found a few bone fragments, loose vertebrae, and shark's teeth, but nothing that "panned out," meaning nothing headed back into the chalk wall. While hunting fossils, he loved to comment on the plants, animals, weather conditions, and scenery. Gazing out into the distance, he asked, "Can't you just imagine the Northern Cheyenne appearing over the horizon?" and answered his own question with a rhythmic chant sung in a minor key. We hunted for several hours that morning, then decided to break for lunch. He discussed pioneering paleontologists over our meal of hard-boiled eggs, fruit cocktail, and bologna sandwiches. Since we had tramped around in the beds for at least three hours by then, our sweat evaporating in the dry air, no lunch ever tasted better.

In all my talks with my father, I never ceased to appreciate his capacity for wonder, which seemed boundless. It was as if living in such a deserted place had kindled in him the ability to look deeper, to find something where others saw little. In *Great Plains Literature,* Linda Ray Pratt noted the effect of the flatlands on its writers: "Climate and landscape play on the psyche, stirring two conflicting responses to the plains environment. Some find the space isolating and fearful; others find it liberating. ... many longtime residents of the Great Plains speak of the *need* for the open horizon before them."[6] Pop needed this openness. As a side effect of many decades of living there, the region quite simply created an ever-unfolding expanse inside my father's mind.

The Great Plains—a large swath of open land in the middle of the Continental United States running from Canada to Mexico—contained on its western edge the High Plains, a narrower area that was covered with short grasses in far western Kansas and extended to the Colorado Rockies. The Great Plains were the home of numerous Native American tribes, horse- and bison-based cultures, before white settlement. Some of the tribes that lived in western Kansas were the Cheyenne, Arapaho, Omaha, and Pawnee. The tragic end of their cultures has been well-documented: White families were supported by the U.S. Army and government policies during the Indian Wars, which effectively drove Native Americans either into the edges of white frontier society, onto reservations, or to extermination. After the Plains tribes were gone (a fact that engendered in my father a romanticized nostalgia and "white guilt" over their fate), western Kansas turned into farming, ranching, and oil-producing country. But the western counties of the state also contained chalk beds that held Cretaceous fossils, which became one of my father's passions.

Over the years, as Pop grew older, he held papers farther and farther away from his eyes and said, "I need arm extensions!" Without the benefit of an eye doctor's diagnosis, he knew he was farsighted. That was another reason he was able to see a speck and know it was a coyote. Dad's unique plains vision, both literal and figurative, was an inspiration to me. His vision was his unorthodox perspective, an insight into a unique place.

Another type of vision common to the plains, the mirage, most often occurs on the flatlands. Even though the fossil beds offer diverse landforms—buttes, hoodoos, ravines—the farmland portions of the western Kansas counties are completely flat by comparison. There is a wide-open view because the earth and sky meet as two featureless planes at the horizon. Driving the highways that connect counties and towns, such as State Highway 96 connecting Leoti and Tribune, a traveler can sometimes see a mirage that shows the next town's grain elevators shimmering on the highway, even though there's no conceivable way they could be so close.

By 1984, I had finished my thesis (whose primary acknowledgments were for my father) and received my master's degree from the University of Kansas. I was living in the central Kansas town of Stafford with my first husband, and during my brief time there gave birth to my daughter, on Valentine's Day. Two months later, my sister Chris traveled from Hays to meet her new niece. I was anxious to get out of the house, so we decided to take the baby and go explore some antique shops in nearby Great Bend.

In a large shop on Highway 96, while gazing into a glass case of human curios, I saw a partial skeleton of a fossil fish from the Cretaceous chalk of western Kansas and recognized it as an *Ichthyodectes*. I beckoned Chris over to see it. "Look, that's from the fossil beds. What is it doing in an antique store?" Chris asked the proprietor, "Is this fish from the fossil fields of Logan County?" and I blurted out, "Did you collect it?"

The fossil broker gaped at us in horror and replied, "Oh, no! I couldn't do that! There's an Old Man of the Fossil Beds, and he guards who comes and goes. I would never try to collect something myself—he knows the landowners, and they don't like strangers trespassing."

I looked at Chris, wide-eyed, and then we smiled at each other knowingly. We both knew that the scattered bones of this type of fish were plentiful in the chalk of western Kansas, at least compared to other types of fossils there.

The store owner didn't say how he obtained the fossil; nor did we ask what price he wanted for it. We walked back to the car, marveling that in this shop, one-hundred-fifty miles east of the fossil beds of Logan County, a legend had developed about our father, Marion Bonner. He *was* the Old Man of the Fossil Beds, and he was also *our* Old Man. At that point in his life, at age seventy-three, he was a recognized field paleontologist known inside and outside of scientific circles.

Pop was always an unrestrained character. When I told him the story from the antique shop and commented on his mythic status, he was delighted. "Good!" he exclaimed. "We don't need a bunch of numbskulls going in there and wrecking fossils. They're meant for science, not tourists."

He took pride in being considered the steward of the fossil beds, specifically the outcroppings of Cretaceous Niobrara chalk exposed along the tracts of the Smoky Hill River in Logan and Gove counties. For more than sixty years, our father collected from these beds and contributed a vast number of specimens to museums in North America. No matter what challenges he encountered in his life, paleontology was his bedrock.

As the youngest of his eight children and the one who spent the most time at home with him after my mother passed away, I have always wanted to share the story of my fascinating father with the rest of the world. *Old Man of the Fossil Beds* is that tale, but it is more than a scientific biography. It is also an account of a bygone era in western Kansas and a memoir about the connection between a father and his family. Most of the Old Man's history took place in six rectangular High Plains counties—Logan, Gove, Greeley, Wichita, Scott, and Lane. And integral to the narrative was our Mom, his brilliant, talented wife, and their eight children who grew up in the fossil beds.

Dad shared so many of his thoughts with me during my high school and college years that I instinctively knew how he would react to almost everything. Even though he's been gone for more than three decades, he's still in my head. Not a day goes by that I don't think of something my Old Man said or did, use his expressions, or look at life from his point of view.

He is more a part of my psyche now than when he was alive. This is true to some extent for my siblings as well. Our extremely unconventional upbringing was not perfect, but we thrived in separate ways. And as we age, we appreciate all that our parents gave us and are embracing the "Old Man" in all of us. If Pop could see us now, he would say, "None of you birds are spring chickens anymore."

My Old Man cradled my infant daughter in February 1984, Stafford, Kansas. When her little hand clung to his finger like it was a twig, he proclaimed, "Darwin was right!" Photo by Clare Jane (Bonner) Askey.

Plains vision is contagious. We children were affected by Dad's perceptions of the land, the fossils in the strata below, and the expanse of imagination that was his greatest gift. When you are raised this way, boredom is not an option.

Now that I'm well into my sixties, my memories of my father are coming full circle. I'm experiencing some of the feelings he must have held in grandparent-hood. But in my younger days, in the fishbowl of the small town, I grew up simply embarrassed. I remember we were sometimes destitute and felt judged by the townsfolk. While Dad was hellbent on "finishing the job of raising me" in Leoti, some of our differences from normal families mortified the adolescent me. I viewed as abnormal our poverty, his single-fatherhood, his age, his drinking, his eccentricities. But as a parent, Pop was never abusive to me—always gentle and sympathetic even when I made bad decisions. The most extreme thing he did was tease me so that I didn't take my teenage catastrophes too seriously. "Thousands killed and many injured," he would say. "You'll be all right." I took the second part of that quote to heart.

After surviving the Depression, the Dust Bowl, and the World War II years, Dad and Mom taught their children a sense of gratitude. They also gave my older brothers and sisters a somewhat middle-class upbringing. Their post-War prosperity was evidenced in two new homes in Leoti and bolstered by a healthy income from wheat farming and the town's movie theater. But by the time the three youngest children, Chuck, Dana, and I, were teenagers, the prosperity was long gone. We lived well below the poverty line, but no matter our privations, our stubborn, nonconformist Dad kept us going, never giving up on himself or us.

He came of age during Prohibition and dealt with grief and the strain of single parenthood by drinking. This was another awkwardness for me, but we all handled our father's alcohol use differently. I mainly coped by playing along and joking with him. I carried a measure of shame about it and was paranoid that the whole town knew all our business. However, decades later, I realize they probably didn't know as much as I feared. Now I fully embrace my past, realizing that our lack of money and Dad's reliance on alcohol was counter-balanced with emotional support and intellectual richness.

Like many children who grow up poor and experience trauma, I spent many years wanting to forget and escape the past. My childhood trauma was my moth-er's early death when she was fifty and I was nine. How lucky I was, though, to have a father like no other.

Kansas Immigrants
1910s–20s

MOST OF THE KNOWLEDGE I have of my Pop's family history came directly from his mouth during the time I spent with him, but my seven brothers and sisters have also shared their experiences of life with our father from the late 1930s to his death in 1992. The anecdotes and memories Dad told us over the years carried different inflections depending on his listeners.

The larger-than-life character who would become the Old Man of the Fossil Beds started out small and spunky. He was born in Imperial, Nebraska, the county seat of Chase County, on May 5, 1911. For three generations, the Bonner family lived on the Great Plains, first in Nebraska, then in Kansas. I can see Dad in my mind's eye—a cheerful lad, running all over Imperial with his cousins. The biggest drama he experienced there happened one day when he and his cousin Cye were "monkeying around" at the windmill behind the Bonner home.

Dad was an inquisitive, sandy-haired four year old. He and Cye were playing a game that involved sticking a finger in the pumping mechanism of a windmill during the upstroke and pulling the finger out before the downstroke. Four-year-old Pop hesitated an instant too long, and the downstroke neatly sliced off his finger in the middle of the second joint.

"Get Mama!" he screamed at Cye, whose eyes were wide with horror. The injured youngster's own blue eyes streamed with tears as he plopped to the ground, and soon his mother came and gathered him up. There was nothing to be done for the severed piece of his small finger. Grandma Bonner bound the wound, kept it clean, and it healed well. What was left was a "pointer" stub.

In future years, when grandchildren asked the Old Man, "What happened to your finger, Grampa?" he would answer with a wide smile, "Wore it off pointing at scenery," or "Steamboat run over it!" He didn't want to traumatize any youngster with the tale, but the details came out when we were older. Fortunately, the digit—what was left of it—was on his left hand and he was right-handed. Dad used that middle finger and stub together to point out things, fret a guitar, and give other drivers the "howdy" wave. Unfazed by the incident, he remained an enthusiastic, fun-loving fellow with an outsized curiosity for the rest of his life.

My father was the seventh and youngest child of Orville Wesley Bonner and Viletta (Markley) Bonner. Before him came three sisters—Veda, Velma, and Helen—and three brothers—Eldredge, Virgil (who died young), and Jennings. In the generation before them, the Bonner family migrated to western Nebraska after the end of the Civil War to make a living off the land. There are records of homestead claims in Hamilton and Chase counties by Dad's "Grand-Pappy," Lewis Charles Bonner, a Civil War veteran who fought for the Union Army.

The Bonners were drawn to western Nebraska because of the Homestead Act of 1862. But as it turned out, there were too many of Lewis C. Bonner's sons trying to make a go of it in Chase County. After attempting to work with his oldest brother, my grandfather, Orville Wesley, decided to take his family further south. He wanted to start fresh, with new land to farm in far western Kansas.

Dad and his cousins pose with their Union Army veteran grandfather, Lewis C. Bonner, in Nebraska, circa 1914. Left to right: Emerson, Cye, Lewis, Marion, and Jennings.

Dad's family moved to Wichita County, Kansas, in the fall of 1919, when he was eight years old. The distance from Imperial, Nebraska, straight south to Leoti was one hundred and eighty miles. Much of their journey was over dirt roads or paths that had been newly etched in the buffalo grass prairie. (Indeed, the north-south Kansas Highway 25 in western Kansas was not paved until 1926.) The small caravan was made up of wagons to haul their household goods, a Model T Ford, and a large family sedan, the Pathfinder, a car that was only produced from 1910 through 1917.

Portions of western Kansas were wild and uninhabited. The cars drawing primitive trailers sometimes went off-road, navigating over short-grass prairie. The Bonners had to scout where the lowest or driest points of east-west running creeks were in order to cross them and continue their southward journey. In some flat stretches of the counties they passed through, the brown alluvial soil was being plowed up and readied for planting with corn or wheat.

In later years, Dad was fond of telling us, with bravado, "I came to Kansas with a gun on my hip." As the family passed through the town of Oakley in the northwest part of Logan County, they encountered some rough old cowhands who thought the gun-toting child was comical. But according to our father, the Colt 45 pistol in his holster was very real. In that era, to allow a precocious eight-year-old boy to carry a firearm probably was not unusual.

It was the tail end of the Wild West. The vast herds of American bison had disappeared, as had the Plains Indian tribes. As noted author Wallace Stegner has written, the end of the Indigenous cultures of the plains was both tragic and ironic. Of course, Native American people thrived on their own before European settlement, but Stegner described how the immigrants' progress through the continent created disaster. He posited in *Wolf Willow*, a book about his childhood on the northern plains, that "the white man literally created the culture of the Plains Indians by bringing them the horse and the gun; and just as surely, by conquest, disease, trade rum, and the destruction of the buffalo, he doomed what he had created."[1]

The predominantly European American immigrants to the plains states took over the vast grasslands, traveling west, eager for a more promising livelihood. Although Dad was the youngest of Orville and Viletta's children, he was not the youngest of the group of Nebraskans; his niece Janice (pronounced "Ja-*neese*") was also in the caravan of hardy travelers. Janice, the daughter of his sister Veda, was five years younger than Dad, and he treated her like a little sister.

By the early part of the 1900s, as the Great Plains were becoming settled, small farming communities like Leoti dotted the land. North of the farmlands of Wichita, Greeley, and Scott counties were tracts of land filled with rocky outcroppings and desert plants such as yucca and prickly pear cactus.

Dad and his niece in Imperial, 1918. He points toward the camera with his intact index finger.

Parts of Wallace, Gove, and Logan counties were less arable, so here the prairie was still unplowed, looking much as it did before settlement. The pastures containing chalk beds stretched for miles along the Smoky Hill River in Logan and Gove counties. Dad glimpsed these chalk bluffs for the first time on the Bonners' migration south and had no idea then that the chalk bluffs would become his second home.

Wichita County, nowhere close to the city of Wichita in Sedgwick County, sits midway between Nebraska and Oklahoma and is the second county east of Colorado. Its county seat, Leoti, was a settlement of approximately four hundred people when the Bonner family arrived. The small community had become the county seat after a protracted and bloody feud between its citizens and the people of Coronado, three miles due east. The infamous County Seat Fight of 1887 between Leoti and Coronado attracted lawmen from as far away as Dodge City, including Wyatt Earp, Bat Masterson, Bill Tilghman, and others. Some claim it was the bloodiest county seat fight in the United States.[2] Four men were killed and three were severely wounded. In *West of Wichita,* Kansas historian Craig Miner mentioned that the event gained national notice; *The New York Times* ran a front page story in March 1887 about the "Cowboy War" in Wichita County.[3] As new immigrants, the Bonner family had undoubtedly heard about the County Seat Fight that took place more than three decades before their arrival.

One of the first things O. W. Bonner and his sons did on their land southwest of Leoti was to "break the prairie" using a Rumely tractor in order to plant first corn, then winter wheat. (Named after German immigrant Meinrad Rumely, the Advance-Rumely company was one of the first mass producers of tractors and threshing machines, which revolutionized farming.) The Nebraskans were more accustomed to corn than wheat, and the Bonners' first corn crop, in 1920, was successful because it was a wet year. Their two quarter sections of land were located directly north of Leoti near Ladder Creek, and the Bonners farmed them until they sold their acreage in 1924. O. W. Bonner bought the land from the Piper family, so until their neighbors learned their names, they were called "the people at the Piper place." According to a record in the Wichita County Register of Deeds' office, the parcels of land that "O. W. Bonner and wife" later sold in 1924 were the southeast quarter of Section 16, Range 17, Township 37 and the northeast quarter of Section 21, Range 17, Township 37.

All over western Kansas, newly arrived farmers cut through the buffalo grass sod to plow into the rich topsoil. In addition to turning over their own land, Dad and Jennings used the Bonners' machinery to provide this service to other farmers. One such project in 1925 involved breaking out one hundred and sixty acres (a quarter section) of land for the Kreutzer family near Marienthal, "by Skeet and Jennings Bonner (of Leoti) using a large Rumley [*sic*] tractor and mold board sod plow."[4] A picture in the Bonner photo albums shows Dad as a lanky youngster with his parents, Veda, Jennings, and Janice, posing on the "first John Deere [tractor] to come to Wichita County, Kansas."

Winter wheat was well-suited to the western Kansas climate, and by 1910, wheat had surpassed corn as the dominant crop in the state. Because it is planted in fall and harvested in spring, winter wheat takes advantage of the limited moisture on the semi-arid High Plains, sprouting underground and watered by snowfall in winter but ready to harvest before the worst heat of the summer.

Dad's parents O. W. Bonner, top, and Viletta Bonner, bottom middle, pose with members of the Bonner family before a wheat harvest with an early thresher.

It would take several seasons of farming before the newly arrived immigrants in all parts of western Kansas were hit with a sobering reality—rain was scarce out here. Being at the mercy of dry seasons and extreme variations in temperature meant dryland farming was a risky proposition. Dad, along with the rest of his family, learned from an early age to obsess about the weather. But the Bonners also viewed the weather with forbearance. They would deal with whatever happened.

Dad called his father "Pappy." In addition to farming, Pappy kept his eye out for new business opportunities in the small town. He dabbled in real estate and land sales, and, in 1926, purchased an existing motion picture business. O. W., Viletta, and their children managed both Bonner enterprises in Wichita County—wheat farming and the "show business." They ran the movie theater initially in the depot building near the railroad track, four blocks north of Main Street (Highway 96). At first the shows were black-and-white silents, dramas starring the likes of Rudolph Valentino, Charlie Chaplin, and Pola Negri. The Bonners ran silent movies (which had accompanying music either from a Victrola or a piano) until the late 1920s.

My oldest brother, Orv, frequently cites an example of how "show business" and entertaining came naturally to Pop. Our aunts reported that back in Imperial, little Marion would get up in front of the audience of a movie theater owned by friends of the Bonners. Before the show, the natural pitchman in Dad came out, and he sang loudly, "Popcorn! Popcorn! Who wants popcorn? You ... you ... you!" In the last part of the ditty, he dramatically pointed to people in the audience.

One Leoti resident remembered the citizens of Wichita County coming to town on movie nights: "On Saturday night ... many of them would go to the show which was in the old Santa Fe Depot Building. ... The first movie I saw there was *Laugh, Clown, Laugh* [a 1928 drama starring Lon Chaney]. Those were the silent movie days, and Mrs. Bonner would play the piano for entertainment before the show started."[5]

In addition to the farm acreage north of Ladder Creek, the Bonner family also bought a small house in Leoti, just north of the railroad tracks. They had enjoyed electric lights in their home in Nebraska and missed them in Leoti, so Pappy pushed for electricity to be installed in the town. The oldest Bonner son, Eldredge, graded roads in his job with Wichita County and also helped maintain the generators that sent electricity throughout the municipality.

As a teenager, assisting with the family's farming and theater enterprises were thrilling jobs for Pop, a boy with a high spirit and sense of adventure. In this simpler era, the luckiest boys were those who entertained themselves with movies, books, hunting, fishing, and games when their hard chores and schoolwork were done. I remember these stories with a sense of jealousy—in the times Dad described, and even in later years when he and Mom were raising the kids, boys did all the fun stuff. Yes, they worked hard, but they didn't have the more restrictive roles that girls did. Wife, mother, or teacher were about the only occupations to which girls could aspire.

Pop often told us about how he and his brother Jennings were able to catch fish from White Woman Creek using a strip of red rag. The White Woman, south of Leoti, and Beaver (now Ladder) Creek, north of town, were teeming with fish in those days. On a typical fishing day, they were able to land catfish, sunfish (perch), and bluegill. Today, primarily due to the depletion of the Ogallala aquifer, a prehistoric water source under the High Plains, the Ladder and White Woman creeks are considered dry creeks, with water in them only after torrential rains. Before it dried up, Dad and Uncle Jennings also fished Chalk Creek, located just inside Logan County.

Fur-bearing animals were also plentiful. The Bonner boys trapped beavers, badgers, skunks, and raccoons, in an era when college kids wore "coonskin" coats, and badger and beaver hats were in vogue. Indeed, beavers were so plentiful that the mascot for Leoti's rival high school to the east, Scott City, was the Beavers. Leoti's mascot was the "Indian." Both Indigenous, both gone.

In small towns of that era, citizens called town characters insensitive names like "Shorty," "Blackie," or "Gimp" if a person had a limp. Inheriting a large dose of cowboy culture, youths teased each other mercilessly. Dad's appellation wasn't too bad. He was called "Skeet" because of the way he buzzed around like a mosquito, a.k.a. "skeeter." It stayed with him for life and distinguished him from the other Bonner boys.

One day in August 1923, the wind started kicking up dirt from roads and fallow fields around Leoti. Twelve-year-old Skeet Bonner had a job as a shoeshine boy in a barbershop owned by B. Frank "Plute" Clayton. Dad worked for Roy Price, a barber who leased part of the building. It was the middle of the day, and the storm came roaring into Leoti so quickly that most people had little time to take cover or go home. Roy saw how bad it was getting and told his astonished worker, "Boy, get under the barber chair!" In later years, Pop said that clinging to the chair, which was bolted to the floor, kept him safe and protected him from falling debris.

The tornado took the roof off the barbershop and several other downtown businesses and flattened whole buildings farther south along Main Street. The shop's customers and hired help, including Dad, were not hurt. He remembered that in the aftermath of the tornado, a young man came to the shop's door, asking for assistance. Plute Clayton, the building's landlord, wouldn't let him in. "Thousands killed, and many injured," Plute said laconically, as if quoting a newspaper headline. "You'll be all right," he concluded. The injured boy stared, unbelieving, but by then, help had arrived from outside the barbershop. The "thousands killed" quote became one of Dad's favorite one-liners, used when people overreacted in a crisis.

Plute Clayton, a Chicagoan, was one of Leoti's earliest settlers. He showed up in Leoti wearing a silk top hat. He was the first barber in town, then later was owner of a mercantile store and a mortician selling caskets. The early Leoti children were fascinated by this colorful character. They got their candy from his store and were elated with the generous portions a penny could buy.

Dad's oldest sister, our Aunt Veda, claimed that Jennings and Eldredge missed the storm because they were hauling furniture from Nebraska. Veda and Velma were outside the Bonner home north of the railroad tracks and started sprinting to the house but didn't get there in time. O. W. was clinging to a pole near the house and told the two girls to hang on tightly to him. Fortunately, no debris hit them in the long minutes when they clung to the pole and to each other.

The scale measuring the severity of tornadoes did not exist in 1923, but the tornado was doubtless a terrifying event. The storm started on the west side of Leoti, missed the high school, and demolished buildings as it made its way to the old Methodist church, which was hit and damaged. In addition to the buildings on Main Street, the tornado flattened the brick grade school.

For the rest of his life, Pop held a healthy respect for tornadoes. If storms were coming and he was out in a car, he got in the habit of trying to outrun them. It worked most of the time because on the plains, during the day, you can see the direction of their movement. Many of the windstorms in western Kansas move from the southwest to the northeast, so they are somewhat predictable. But not always.

Fish Catches Boy

1920s

SKEET WAS A LATE-BLOOMER, undersized until his twentieth year, when he reached his full height of five feet, ten inches. When he entered Leoti High School[1] in 1925, the slight, wiry boy started playing guard for the basketball team and was an excellent free-throw shooter. In one contest, he won a pewter basketball-shaped trophy for hitting fifty free throws without a miss. In later times, Dad would occasionally emerge from his room, brandishing the trophy, and proudly exclaim, "Fifty straight!"

In high school, he enjoyed playing pranks on his friends, telling tall tales, and kidding around with the girls. But he was also captivated by science, particularly the idea of geologic time and the existence of prehistoric animals that lived on Earth millions of years ago. Dad took a wide variety of courses, including music appreciation, commercial law, Shakespeare, and Latin, but it was science—particularly geology and paleontology—that he loved the most and that had the most impact on his life.

At that time, spurred by the discovery of dinosaurs, geology and paleontology began to enter the popular imagination and school curricula. Within geology, which deals with the history of the Earth as recorded in rocks and minerals, is the subset called stratigraphy. Stratigraphy analyzes the origin, composition, distribution, and succession of rock strata. Two early scientists, considered the "fathers" of geology, were the Scots James Hutton, author of *The Theory of the Earth* (1788), and Charles Lyell, author of *Principles of Geology* (1834). An Englishman who singlehandedly created the first geologic map was William Smith, who also published early works on stratigraphy.[2]

Paleontology was born in Europe in the 1700s, growing out of the French scientist Georges Cuvier's studies of comparative anatomy (comparison of fossil bones to modern ones). Cuvier's work, *Researches on Fossil Bones,* published in 1812, is considered the basis of modern vertebrate paleontology.[3] The word "fossil" originated in the 1650s and meant anything dug up or unearthed. Paleontology deals with life from past geologic periods as known from fossil remains; its branches include vertebrate and invertebrate paleontology. It focuses on life that existed before the current geological epoch, the Holocene, or post-Pleistocene Epoch (roughly 11,700 years ago).

The study of ancient life exploded in the U.S. in the late 1800s and early 1900s when fossil hunters, exploring the terrain of the West, began discovering gigantic dinosaurs in mountainous areas that bordered the Great Plains. Dinosaurs were not found in the fossil badlands of the Great Plains themselves, which had been covered by an interior seaway during the time these beasts roamed. As Kansas paleontologist Michael Everhart points out in *Oceans of Kansas,* dinosaur finds in Kansas are extremely rare. Over many decades, the Smoky Hill Chalk has produced only a handful of hadrosaurs and nodosaurs, and it is thought those dinosaurs died near the water and washed into or were dragged into the sea, which may explain why they were found in the Cretaceous marine environment.[4]

What Kansas became famous for was ancient marine life. The Kansas chalk produced fossil reptiles (pterosaurs, mosasaurs, plesiosaurs, and turtles); bony and cartilaginous fish and sharks; toothed birds (flightless swimmers and gull-like flyers), and invertebrates (bivalves, oysters, squids, and crinoids).

When he was fourteen, Dad met the man who would become his first mentor, a brilliant science teacher named Arthur A. Wedel. Whenever Pop talked about Mr. Wedel, it was in reverent tones. From a Swiss Mennonite family based in Moundridge, Kansas, Wedel was working toward his doctorate in geology, which he would earn from Cornell in 1930. In 1925, Leoti was the last of four Kansas high schools where he taught science.[5] Mr. Wedel told the Leoti students that he was lured to western Kansas by the Cretaceous chalk, and Dad soon learned why.

The time span of the Cretaceous Period was between 145.5 and 65.5 million years ago, although scientists in the 1920s could only guess it was "millions of years." The Cretaceous was the third and final period of the Mesozoic Era, and the longest. It followed the Triassic and Jurassic periods and ended with the extinction of the dinosaurs. The name "Cretaceous" derives from "creta," Latin for chalk. The Western Interior Seaway, existing for millennia, divided two land masses, Laramidia on the west and Appalachia on the east. During the Cretaceous, the coastline shifted multiple times, and this caused the marine animals to alter and adapt. The changing coastline is one of the many factors that caused the evolution and development of so many species of marine animals and the diversity of the dinosaurs living on land on either side of the seaway.[6]

Educating the public about the geological significance of the Cretaceous chalk dates as far back as the founding thinkers of geology, Hutton and Lyell. In addition, Thomas Huxley, a comparative anatomist and evolutionary theorist, studied the formations of the Earth. Huxley contributed an essay to *Macmillan's Mag-*

azine called "On a Piece of Chalk" in 1868. To help his audience comprehend the age of the Cretaceous chalk in England, famous for areas such as the White Cliffs of Dover, Huxley wrote:

> The area on which we stand has been first sea and then land, for at least four alternations. ... During the chalk period, or "cretaceous epoch," not one of the present great physical features of the globe was in existence. Our great mountain ranges, Pyrenees, Alps, Himalayas, Andes, [to which American scientists would add the Rockies] have all been upheaved since the chalk was deposited, and the cretaceous sea flowed over the sites of Sinai and Ararat. All this is certain, because rocks of cretaceous, or still later, date have shared in the elevatory movements which gave rise to these mountain chains.[7]

Focusing earnestly on his students, Mr. Wedel educated them in the scientific theories of Hutton, Lyell, and Huxley. He told his students that fossilization is the process whereby the organic material of ancient bones slowly leaches away over time and is replaced by minerals that harden the bones, preserving the animals where they died. He also told them he was keen to explore the chalk formations in Logan County, north of Wichita County. Wedel was aware that early paleontologists like Benjamin F. Mudge, Edward D. Cope, Othniel C. Marsh, and Charles H. Sternberg had discovered major fossils in the Kansas chalk, primarily in the Smoky Hill outcroppings of Logan and Gove counties.

One day, Mr. Wedel informed his science classes that he was planning to take a field trip to the chalk beds the next Saturday and asked if any students wanted to go with him. Six boys, including Dad and his brother Jennings, signed up for the excursion. That was a lucky number because Mr. Wedel's Ford Model T could hold only seven people.

On a balmy Saturday in September, they traveled north of Leoti over a dirt road. Several miles into Logan County, Mr. Wedel turned east for a few miles, then passed into a rough pasture. As the Model T slowly chugged along over bumpy, brittle prairie, the teacher pointed out a large canyon full of chalk formations jutting out of the valley. The colorful stratigraphy was yellow to orangish chalk for the top two thirds, and bluish gray chalk for the bottom third. Fully exposed, free-standing spires—two stories tall—rose up out of the valley, darkened by weather and lichens.

Freshman Skeet Bonner (top row, middle) with fellow high school students in the fossil beds in 1925. His older brother Jennings, a junior, is at far right. Photo by Arthur Wedel.

As they pulled into the pasture and looked at the bluffs, he told the boys, "These strata were deposited very slowly over eons of time, when a prehistoric sea covered Kansas and flowed over the middle of the continent. That sea was the Cretaceous Sea, dating back millions of years.[8] The silty sediments and marine life settled, compressed, and turned into these chalk bluffs."

The boys looked around them in wonder. Skeet Bonner, a freshman and the smallest boy in the group, hopped from the car and darted across the prairie, jumping over small clumps of milkweed. The other boys piled out of Mr. Wedel's Model T and headed down the draw. In front of them, they saw yellow bluffs carved out of the grassy hills.

"If these fossil beds were under the sea, why do they look like this now, Mr. Wedel?" Pop asked his teacher. Wedel said that after the passage of millions of years, the chalk deposits were exposed by wind and water erosion to create the outcroppings of today. He explained that limestone, chalk, and shale are marine deposits. "The silt and microorganisms formed this rock slowly, over millions of years. The chalk itself is actually a fossil," he stated.[9]

Loud, clicking grasshoppers jumped out of the way of Dad's leather shoes as he walked, gazing at the rocks at his feet. He looked back at Mr. Wedel, who was carefully hiking down the gully behind him. The boys ribbed the teacher for taking his time. Then Dad saw, by Mr. Wedel's boot, a dark gray bone sticking out of the chalk. He yelled, "Mr. Wedel! By your foot! It's a dinosaur!"

The teacher stopped and examined the bone. It looked like part of an animal's jaw. He shouted, "Come here, boys, and look at this!" He explained this couldn't be a dinosaur because during the late Cretaceous, those giants lived in the area that was now Colorado and Wyoming, while western Kansas was covered by a large sea. "This bone," he explained, "looks like a part of an ancient fish that lived in the sea at the same time as the dinosaurs." One of the boys spotted a dark shiny triangle a yard away from the fish. "That is a shark's tooth! Maybe that shark fed on this fish," Mr. Wedel theorized.

"What kind of fish is it? Can we dig it out?" Dad's heart raced with anticipation. The fossil lay exposed, flat on the rocky shelf as if dropped there from above. Actually, it was the rain and wind of western Kansas that had exposed the bone. Mr. Wedel and Dad looked closer and carefully removed chalk around the fossil jaw with an awl-type tool. They uncovered small, sharp teeth, showing it was indeed the jawbone of a fossil fish.

Dad was hooked! From that point on, thoughts of ancient fish that lived right where he was standing consumed him. The bones the boys found that day were all scattered parts of fishes. They found more shark's teeth, fish vertebrae, and coprolites, which their teacher said were fossilized fish droppings. The boys hooted at that, but they all agreed that the plentiful coprolites, whiter than the surrounding beds, looked exactly like dung that had turned to chalk. The jaw Mr. Wedel nearly stepped on at the beginning of their hunt, which he said was probably a *Gillicus*, was the highlight of the day. Mr. Wedel gathered the group of boys together and snapped a photograph of them leaning and sitting on a knoll.

Gillicus arcuatus was a ray-finned fish related to *Ichthyodectes* and *Xiphactinus*. It was the smallest of the three, reaching lengths of six feet. *Gillicus* had modest-sized teeth lining its jaws and ate tinier fish by sucking them into its mouth. It possibly also fed on organisms such as plankton. The famous paleontologist Edward Drinker Cope (1840–1897) originally named *Gillicus* as *Xiphactinus arcuatus,* denoting the species (which means "arched") from a type specimen collected in the Kansas chalk in 1875. The name was later amended to *Gillicus arcuatus.* A type specimen, or holotype, is the original specimen that defines a species.

At school the following week, Dad pored over books Mr. Wedel owned that described mosasaurs, plesiosaurs, a large fish that Cope called *Portheus* (now known as *Xiphactinus,* the original name given it by Dr. Joseph Leidy, one of the most influential scientists of the era),[10] *Gillicus*, and smaller fish as well. The famous paleontologists of the 1800s, Edward Drinker Cope and Othniel Charles Marsh (1831–1899), had also found flying animals that had fossilized in the sea's sediments: *Pteranodon* and ancient birds like *Hesperornis* and *Ichthyornis*. "All of these animals are in the rock just waiting to be discovered," Dad thought, amazed. That was the same notion most paleontologists had upon viewing the chalk outcroppings along the Smoky Hill riverbed. The Smoky Hill River didn't emerge as a "river" unless hit by a strong gully-washer, but after the rain, any accumulation of water instantly sank into the dry land.

The abundance of fossils led to a flurry of collection between two intense competitors, Cope and Marsh, the early paleontologists known for the "Bone Wars." They began their feud when collecting dinosaurs, and the chalk beds of western Kansas were another area they vied over. Initially, Marsh worked to expand the holdings of the Yale Peabody Museum in New Haven, Connecticut, while Cope sent most of his discoveries to Joseph Leidy in the Academy of Natural Sciences in Philadelphia and to the American Museum of Natural History in New York.[11]

Cope was born in 1840 to a Quaker family. He was largely self-taught but took a few courses at the University of Pennsylvania, where he learned about comparative anatomy from Dr. Leidy. Cope published voluminously, and by the time of his death, had "described 1,115 of the 3,200 species of vertebrate fossils then known from North America."[12] Marsh, in contrast, was college educated, with a master's degree from Yale. He studied extensively in Germany, then the mecca for science. (In fact, Cope and Marsh met in Germany and were friends there during the Civil War. Cope had been sent there to avoid the draft.) Favoring complete, dramatic skeletons, Marsh described fewer "new" species than Cope did but shared his passion for acquiring fossils. Both scientists employed teams of fossil hunters and excavators traveling all over the western states. During their long careers, they never stopped competing to see who could collect the most fossils for their institutions. While the Bone Wars embarrassed the paleontological community, the fierce competition yielded a tremendous number of scientific discoveries and publications.

Driven by the public's fascination with fossils and the sheer numbers of specimens being collected, natural history museums emerged across the country. Parents wanted to be able to take their children to these museums, built in major cities and often connected to universities, for an educational pastime. In Kansas, the collecting boom expanded two main natural history museums, at the University of Kansas in Lawrence and at Fort Hays State University in Hays.[13] The museum in Hays, first established in 1914, now houses over 100,000 square feet of specimens. The University of Kansas Museum of Natural History, established in 1890 and now called the KU Biodiversity Institute and Natural History Museum, has nearly six hundred type specimens in vertebrate paleontology alone.[14] George Fryer Sternberg, one of Charles H. Sternberg's sons, was a driving force behind procuring fossils for Hays, whereas Samuel Wendell Williston was a prime mover who expanded acquisitions in the Lawrence museum.

Studying the large science books borrowed from Mr. Wedel, Dad created a forty-four-page term paper called "Cretaceous Fishes," which included drawings of specimens he had found in the chalk during the years Mr. Wedel was his teacher. Some of the names of genuses have changed since the 1920s, but when he drafted his paper, he copied by hand Cope's drawings of *Saurodon, Empo, Anogmius, Gillicus, Pachyrhizodus, Enchodus, Protosphyraena, Squalicorax, Isurus, Ichthyodectes,* and other prehistoric fish genuses.[15] Genus names were etched in his brain so that when he went out in the field, he was able to identify the marine animals as soon as he spotted them. Some of the books he consulted were Cope's *The Vertebrata of the Cretaceous Formations of the West*[16] and two volumes from the University Geological Survey of Kansas Paleontology series: *Teleosts,* by Alban Stewart, and *Mosasaurs* by Samuel W. Williston.[17]

In the chalk, with Mr. Wedel's guidance, Dad learned how to carefully dig around an exposed fossil to see if it was complete enough to be collected. There were myriad fish tails and fins to be found, but finding a fish skull or a complete fish was much rarer. The first large fossil he collected, while a high school junior in 1927, was a complete skull of the formidable predator, *Xiphactinus.* Previously called *Portheus* by Cope, *Xiphactinus* ("sword ray") reached lengths of eighteen feet. It preyed on its relatives *Gillicus* and *Ichthyodectes* as well as numerous other smaller fish, sharks, and squids. The largest fish of the era, *Xiphactinus* resembled a huge toothy tarpon.

To collect the large skull, Dad used a technique developed by Charles Sternberg and his sons. He dug around the skull's rock matrix, making a flat, even surface. Then he built a wooden frame that he placed around the fossil's periphery. Next, he filled the frame with plaster of Paris so that it settled and hardened around the chalk containing the skull. After waiting a few hours while it solidified in the arid air, he gingerly dug around and under the cast. Then, very carefully, he pried the cast away from the chalk layer to separate the framed fossil from the rock below it. Finally, he turned over the whole cast containing the fossil. He and a friend then placed the cast on some tires in the back of the Bonner family's farm truck. The tires provided a cushion for the fossil and protected it from being damaged during the bumpy ride. Once home, Dad laid the framed fossil on a large table and painstakingly chipped away the chalk from around the bones, which uncovered the reverse side of the skull.

Word of his discovery spread quickly in the small town. The *Leoti Standard* ran an article with the headline "Leoti Lad Finds Fossil of Huge Proportions." The story described the head of the "genus *Protheus* [*sic*]. In size, the head is a monster ... about 2 ½ feet broad." The story stated that the fish skull was larger than that of the complete *Portheus* [*Xiphactinus*] found earlier by "Prof. Sternberg, the foremost fossil hunter in the world." The *Standard*'s hyperbolic description ended by praising Dad's fish in comparison to Sternberg's in Hays: "Sternberg's fish would just about make fair sized bait for Bonner's huge find."

When the inland sea covered North America, its marine animals reigned supreme. The sea lasted 34 million years, forming one hundred million years ago and finally disappearing some 66 million years ago. Dad learned much later that the end of the Cretaceous is marked by the K-T boundary, a term first used in 1983 that marks a change from the Cretaceous (K for *kreide,* or chalk) and Tertiary (or Paleogene) periods. This boundary delineates one of the largest mass extinctions in Earth's history and signals the end of the dinosaurs (except the ancestors of birds) and most of the marine animals. Elevated levels of iridium (an element found in meteorites) in the K-T boundary gave rise to the Alvarez hypothesis that the impact of a massive asteroid brought on the extinction of dinosaurs and many other species. This impact formed the Chicxulub crater near the Yucatán Peninsula in Mexico.

After first claiming the best fossils for their own institutions, early paleontologists then focused on placing the remainder of their specimens in other museums. Some, such as the Lawrence-based Sternbergs, worked for a time as independent collectors. Charles Sternberg and his sons, Barnum Brown (another famed Kansas-born paleontologist), and others sold fossils to institutions that wanted them. Naturally, Dad wondered whether any museum would be interested in his *Xiphactinus* skull.

In 1929, he tried to place the skull with Handel T. Martin, who was then museum paleontologist at the University of Kansas. In May 1929, six days after getting Dad's letter, Martin typed a reply on KU museum stationery that read, in part:

> For the past thirty years I have worked the fossil beds of Logan-Gove-Trego Graham-Scott, Wallace, and in fact all the Cretaceous beds, the proceeds of these yearly trips are all housed in our Museum here, so you will see that we already have a vast collection of fossils from these beds. In fact, we have the best collection from Western Kansas I think known, which includes many skulls, and

some skeletons of *Portheus*, as well as many duplicates of every species known from these beds, both Fish, Reptile, Pterodactyl, and Birds. ... I am returning the Photographs so you will be able to use them for some other prospect. I think you will find that most of the institutions are already supplied with Kansas Cretaceous Fossils. Before coming to K.U. I collected twelve years for commercial use from these beds, and the Sternberg outfit have been collecting things from there for forty years for sale, so you will see most of the Schools and Museums are stocked pretty well.

Dad wondered why Martin's answer was so discouraging. The quality of his large fish skull was excellent. Even if the museum did not want to display the fossil, wouldn't there be value in having it as a study specimen for comparison to other fish? He didn't yet understand how museums worked; in fact, at that point in his life, he had not visited the natural history museum in Lawrence. If he had known how many fish they had, he might have directed his query to a museum farther from Kansas. Also, the more experience he gained as a fossil hunter, the more he understood that Martin, in his way, was probably dissuading an "amateur" from potentially ruining fossils. It was the beginning of Pop's understanding of a push-pull relationship between the field collector and the museum preparator.

Indeed, Handel T. Martin (1862–1931) did have a tremendous amount of experience in the fossil beds of western Kansas. An English immigrant, he lived in Logan County and began collecting fossils from the Smoky Hill Chalk in 1887. He collected in Kansas for O. C. Marsh, and after some time in the field, went to the American Museum of Natural History. There he learned how to mount the bones of vertebrate fossils for exhibition. He started working at the University of Kansas in 1896 as the chief assistant and preparator for Samuel W. Williston and was curator of fossil vertebrates at the University of Kansas from 1912 until his death in 1931.[18]

Dad kept the large *Xiphactinus* skull for many years and never tried to place it with another museum. Nevertheless, Martin's letter made him determined to prove the man wrong. He found it particularly galling that Martin had said the museum had "every species known." In a way, it awakened in Dad a Cope/Marsh-like competitive spirit. In geologic terms, he reasoned, the fossil beds were only beginning to be tapped, and forty years was nothing in geologic time.

Pop filed Martin's letter away in his desk. It became the bottom page of a folder of lifetime correspondence with museum scientists that grew thicker and thicker with time. Paleontology would become his lifelong hobby and obsession. From his first *Xiphactinus* skull onward, our father spent the next sixty years hunting and collecting in his spare time, doggedly disproving the assertion that the beds had been exhausted and had no more secrets to reveal. Martin's letter probably also planted a seed in Pop whereby he could affirm, in future years, that a field paleontologist was just as capable as an academically trained one.

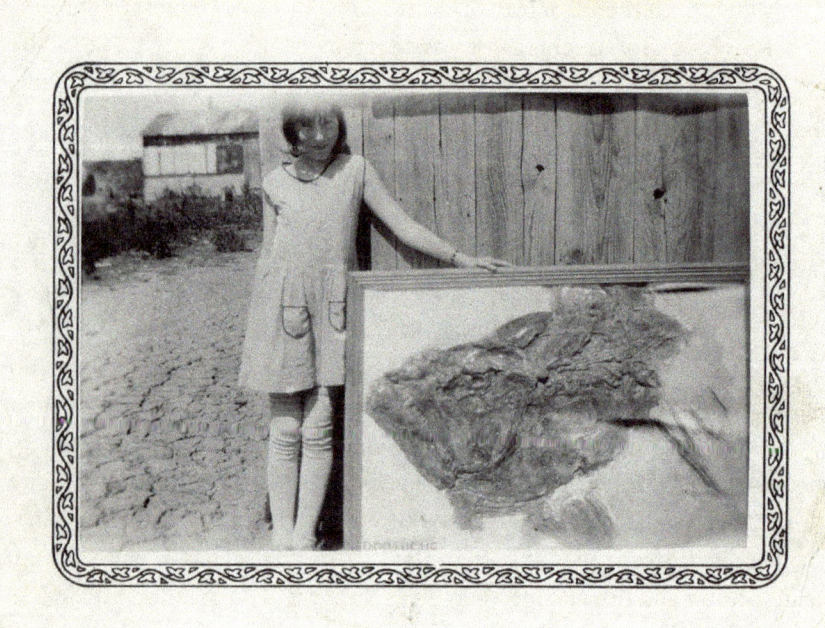

Janice, age eleven, stands beside the large Xiphactinus skull Marion collected in 1927.

Before leaving to finish his doctorate in geology at Cornell, Mr. Wedel left a gift to his students—his Model T. Wedel hoped the science classes would use it to continue hunting fossils, and Dad and his classmates did just that. Mr. Wedel knew he had sparked a thirst for scientific discovery, particularly in the youngest student of his first fossil-hunting trip. He also gave our father a special gift—the large reference books that he had used for his research paper on Cretaceous fishes. These books formed the substratum of a paleontology library that Dad accrued over his lifetime.

Boy Meets Girl
1920s–30s

THE STORY OF MARION "Skeet" Bonner, the Leoti boy who would become my father, begins in the 1920s, the decade that Dad always referred to—with a touch of nostalgia in his voice—as the Roaring Twenties, emphasis on the "roaring." During this decade and the one that followed, Dad continued the pursuit of fossils in this spare time. In 1927, he would meet the scientist who would influence him more than any other: George F. Sternberg. Pop's first mentor, Arthur Wedel, fortuitously connected him to his next mentor by way of a loaner car.

Driving Mr. Wedel's Model T, Pop and two of his high school friends made the one-hundred-forty-mile trip to the new natural history museum at Fort Hays State University in 1927. Mr. Wedel had told them about the museum, so they wanted to see its fossils and meet George Sternberg, of the famous fossil-collecting family. Sternberg had lived in Oakley, in Logan County, before becoming the first paleontologist at Fort Hays State. He was well known as an expert collector with firsthand knowledge of the Cretaceous chalk.

Sternberg was in his lab at the museum when they arrived that morning. He was well dressed, sporting a coffee-colored three-piece suit and hat, which impressed the ragtag visitors. "A study in brown," Dad mused. In talking with the renowned paleontologist, he realized that Sternberg loved imparting knowledge to students. Sternberg encouraged the boys in their fossil-hunting endeavors, and told them, "If you find something good, let me know, so I can collect it for the museum."

Dad was supremely interested in a large, complete *Xiphactinus* that Sternberg had collected and that was now on exhibit. Pop wanted to compare it to the *Xiphactinus* skull he himself had collected, so he bought a four-by-eight-inch photograph from the museum gift shop for the then-dear price of twenty-five cents. The photo shows a five-foot-four George Sternberg standing beside the huge fossil fish, and Dad kept it for the rest of his life.

A Fort Hays museum–issued photograph of George F. Sternberg standing beside a complete Xiphactinus. Dad bought this photo at the museum for a quarter in late 1927, after he collected his first fossil skull.

The Roaring Twenties were a time of prosperity nationally and locally. In western Kansas, farmers responded to the high demand for wheat during the World War I years, producing bumper crops. The farmers' prosperity continued beyond the Great War. Across the country, jazz music and the flapper lifestyle blossomed, women gained the right to vote in 1920, and to get around the Prohibition era—which lasted from 1920 through 1933—people made their own "rotgut" liquor and "home brew," which lubricated parties, card games, and barn dances.

Western Kansas had its own version of the "crazy years," when girls wore shorter, tighter dresses and boys wore baggy pants and long fur coats. At area barn dances, besides the traditional reels and square dances, dancers enjoyed cutting a rug to vaudeville tunes and popular numbers, rushing to the floor to do the Charleston, the Varsity Drag, and the Collegiate. Bootleggers around Wichita County supplied partiers with alcohol. It was rumored that Plute Clayton sold it in his store, under the counter. During one investigation into illicit booze, he famously said to a judge, "No, I ain't never sold no whiskey, and I ain't gonna sell no more."

After graduation, Pop dated a few of the town's lovely ladies and got involved with a band that played at the local barn dances. He was the drummer for a combo called the Whoopie Makers—quite a shocking name for the time. And Pop no doubt lived up to the name. He was a high-spirited bachelor then, fitting right in with the times. An ardent musician, he never stopped being a performer. Pop was always a flirt, but he actually *listened* to members of the opposite sex. He had a knack for finding out what mattered to them. If a young woman happened to be shy, he would simply entertain her.

Being in the "show business" meant the Bonner family had access to the most up-to-date tunes. Dad's sisters Veda and Helen bought the sheet music of many popular ditties. I remember Dad and Jennings, two strong tenors, harmonizing, barbershop-quartet style, on this song that came out in 1925:

> Angry, please don't be angry, 'cause I was only teasing you.
> I wouldn't even let you think of leavin'—
> Just because I love you true.
>
> Just because I took a look at somebody else,
> That's no reason you should put poor me on the shelf,
> Angry, please don't be angry, 'cause I was only teasin' you ...
> (*bass line finish*) only teasin' youuu.[1]

From 1956 through the 1970s, when Dad sat down at the piano, or when I played ragtime music on the ivories, he'd remember the Whoopie Maker gigs. He would rhapsodize about how the crowd would yell to the piano player, "Play the Maple Leaf!" He also reminisced about how the band would sneak out to the fields and find a designated badger hole where home-distilled whiskey was hidden, then would down a mason jar of the stuff before going back in to perform popular songs like "Somebody Stole My Gal," "You're Nobody 'til Somebody Loves You," and "Toot, Toot, Tootsie, Goodbye," all with a touch of Scotch-Irish country roots.

Dad brought his trap set to the gigs. In addition to a piano and his drums, the group had a fiddle, guitar, and sometimes a clarinet or saxophone. The band's repertoire was transitional. They focused on the new rhythms of popular music but kept a few traditional square dances in the mix. They frequented venues in four western Kansas counties: in Wichita County at Wolfenbarger's Beach and Kleymann's Barn; in Logan County at Hillary's Barn and Elkader; in Wallace County at Madigan's Barn; and in Scott County at Beaver Beach.[2]

Dad's drumming rhythm was straight four-four time, but he kept a strong beat and it came through in his later piano playing, when he pounded out by ear the songs the Whoopie Makers had played. He banged at it like the percussion instrument it is, stomping his foot to keep time while he "chorded" and sang the melody, or fiddle part. One of his favorites, the schottische, was a slow polka number he would play in later years with his children, Dad handling the bass and percussion parts on the lower end of the piano and the child picking out the melody on the upper end. Another standard fiddle tune was "The Devil's

Dream," which he also hummed when bouncing children on his knee. Pop's singing and proclamations punctuated our duets. Sitting beside him on the piano bench cranking out Roaring Twenties numbers was one of my favorite pastimes.

Enter Margaret. In 1932, the year Dad turned twenty-one, my maternal grandmother, Clare Doyle Berg, moved to Leoti from Westport, a neighborhood of Kansas City, Missouri, with her two high-school age children, Margaret Christine and Ralph Hartman Berg. Margaret and Ralph were twenty-two months apart in age. The main reason for their move to western Kansas was so Clare could care for her aging parents. Her father, our Mom's grandfather, was the first doctor in Wichita County, Adolphus Merry Doyle. Dr. Doyle and his wife, Margaret (for whom Mom was named), had homesteaded a farm a few miles west of town in 1885. Grandmother Berg's parents didn't live much longer after her arrival in Leoti. Adolphus died later in 1932 at age ninety, and her mother passed a year later, age eighty-nine, in 1933.

The Bergs' background differed in two major ways from that of the Bonners. First, their early years were urban, not rural. Moving from a large city to a small town undoubtedly required some adjustments. Second, one generation back, on the Bonner side, Dad's Grandpappy, Lewis C. Bonner, served in the Union Army during and after the Civil War. Mom's grandfathers, on the other hand, were on both sides of the conflict. Her maternal grandfather, Dr. Doyle, a Missourian, was a soldier on the Confederate side. Her paternal grandfather, Hartman Berg, served on the Union side, in the 209th Regiment of the Pennsylvania Volunteer Infantry. My brother Steve researched our Berg lineage and discovered that Hartman was shot through the chest at Fort Stedman, Virginia, and miraculously survived. Had the ball been an inch closer to his heart, none of the Berg children—or us—would have existed.

Dr. Doyle was lured to Wichita County after the Civil War by free land, where he claimed the standard one-hundred and sixty acres, or a quarter section, under the Homestead Act. He was cited as one of the county's first homesteaders in an item in the *Leoti Standard*: "Among the first settlers were ... a couple of Missourians, P. P. Shearmire and Dr. A. M. Doyle. These folks filed in August and September 1885, and all got land adjoining the town site of Leoti, or nearly so." Dr. Doyle tended to the wounded men who took part in the county seat fight between Leoti and Coronado in 1887. He removed the leg of one of the combatants, Emmitt Denning, with an amputation saw. Dr. Doyle's primitive saw is now in Leoti's Museum of the Great Plains, along with his black leather medical bag.

Grandmother Berg was fond of telling my older brothers and sisters the story of when her brother, Harold, the youngest of the family, was born. The Doyles—the parents and four daughters, Ethel, Lillian, Clare, and Elsie—were living in a dugout on their homestead. Doc Doyle was out on the land tending to his crops when Clare's mother went into labor. "Go get Papa, and tell him to hurry!" she told her daughter. Clare scampered over the plowed furrows toward her father, repeating to herself, "Hope it's a boy! Hope it's a boy!" Doc Doyle ran back to the dugout with his daughter, where he delivered his son, John Harold Doyle.

The Doyle family, with so many girls, was ready for a son. Traditionally, boys were the preferred gender, not only for the labor they provided, but also to perpetuate the family surname. Looking back, it's amazing how male-centered our immediate ancestors and our family were. It was baked in to farm life, Old West culture, and small-town expectations. Looking through contemporary eyes, I wince at the gender restrictions. Our mother handled her role with such aplomb that we never questioned this until the women's liberation movement of the 1960s and 1970s.

On the Berg side, Mom's father, Ralph Harvey Berg, was born in Johnson County, Kansas, in 1885 and married Clare Doyle in Jackson County (Kansas City), Missouri. The couple lived for a time in Omaha, Nebraska, and Ralph worked as a salesman for Marshall Field's, the Chicago-based retailer. In 1917, he registered for the draft for World War I in Marshalltown, Iowa, which was the birthplace of Margaret Christine, born March 21, 1917. Unfortunately, Ralph Harvey died in October 1918 during the Great Influenza epidemic, three months before his son, Ralph, was born. Clare did not have her children until she was in her mid-thirties. Before marrying Ralph Harvey Berg, she had a career as a secretary, having attended secretarial college in Salina, Kansas.

For Christmas in 1916, three months before Mom's birth, Ralph Harvey gave Clare a delightful present from Marshall Field's: an Eastman Kodak 2C camera that used 116 film. To be able to document the family's events with photographs was indeed a luxury at that time. The original list price for the camera was twenty-one dollars. Grandmother Berg took many photos of her family and then gave the camera to Mom, who kept it through her marriage. In our family's albums, any photos that are four by six inches were taken with the big Kodak. From black and white through color film, which was introduced for the camera in the 1950s, Mom documented the family's growth as well as Dad's fossil discoveries.

Margaret and Ralph Berg graduated from Leoti High School two years after they arrived in town. Ralph, sixteen, had enough credits to graduate the same year as his older sister, which he was eager to do so that he could leave Leoti, find a good city job, and help support his single mom. Ralph wanted Clare to have running water and a bathroom instead of an outhouse and a pump in the back yard. The family had had indoor plumbing in Kansas City, and he strongly believed his mother needed those conveniences.

Four years after she brought her family to Leoti, Clare Berg became the first librarian for Wichita County. The library was situated in the county courthouse, which had been dedicated in 1917. A large, two-story red brick structure with columns in "modern colonial" style, the courthouse was Leoti's dominant structure. Clare's wage as a librarian in 1936 would be twenty-four dollars a month—slightly more than the cost of her 1917 Kodak camera.

My mother first came into my father's world through the dances where the Whoopie Makers were performing. Before leaving Leoti, Ralph frequently took his dates to the local dances. Mom and her friends—those who had access to cars—would sometimes go "stag" to the barn dances while in high school. That is most likely when she first saw Pop in action, but being a shy, tall girl, she merely observed and danced to the lively music. She probably also heard a few choice stories about the young drummer who worked at his parents' picture show.

Starting in 1926, the O. W. Bonner family ran movies in the Santa Fe depot, but a few years later, they leased a building farther south on Main Street across from the post office. Each week Dad would walk north to the depot to pick up the film reels Pappy had ordered for the coming week. The Bonners booked all kinds of movies, including westerns, romances, and pirate adventures. Dad adopted the line, "Aha, me proud beauty!" from one such film and would use it on young women (and, later, to tease his daughters while they were applying makeup).

One day he walked to the station to pick up the new releases for the next run of shows. When he lifted the reels and started making his way back down Main Street, one of his high school friends drove up in a new Model A Roadster convertible. "Do you need a lift back to the show, Skeet?" he asked. Sitting beside the handsome young man in the front seat was a comely dark-haired young lady wearing a fashionable coat with a fur collar. Dad regarded her with interest. He accepted the ride and got in the bench seat of the car. To give him room, the girl, Margaret Berg, scooted over to the middle of the seat between the two young men.

They all made small talk, and Dad remembered feeling the length of his thigh brush against the length of Mom's thigh. This was electrifying. Their legs jostled against each other as the car made its way down the brick street. Smiling and looking intently into her large brown eyes, he blurted, "What's a city girl like you doing in this little town?" Stunned to be addressed in such a forthright manner, she responded, "I'm familiar with Leoti. We moved here to take care of our grandparents. My grandfather was Doc Doyle." His friend pulled up to the curb by the theater and Dad got out, thanking him for the ride. Looking at Margaret intently, he said, "I'm sorry about your granddad. It's been nice meeting you." Then, bidding goodbye to Margaret's boyfriend, he had a wild thought: "Wow, she's a beauty. I'm going after her!"

The path to courting Mom was circuitous because the boy who gave Dad the ride had gifted her with a "promise" watch, and Mom had met him in the Presbyterian Sunday school class, which meant they had shared values. But Dad chatted with her every time he saw the statuesque, olive-skinned young lady. He told her about the fossil beds, his eyes lighting up with excitement. Margaret was enthralled with his geological talk. She was smart too—she shared the valedictorian honors with her brother, Ralph, when they graduated in 1934. Before long, Dad's single-minded pursuit of Mom began to bear fruit. She consented to go on a trip to hunt fossils with him (after she decided to break off the "steady" relationship with the other boy and returned his watch).

Leoti High School senior photo of Margaret Berg, 1934.

Dad was much closer to his mother, Viletta, than to his father. O. W. Bonner strongly preferred spending time with his oldest son, Eldredge, over Jennings and Marion, and this made the youngest son more attuned to our Grandmother Bonner. Dad wanted his mother's approval of this girl he was getting so serious about. Viletta had met some of his previous flapper girlfriends and thought them frivolous. After Dad introduced Mom to Grandma Bonner, he asked her, "What did you think of Margaret?" Viletta replied admiringly, "My, she's a bright one!" That was quite an accolade from his mother, so he knew he was on the right track.

Many of the songs, instructional quotations, and admonishments Dad used as a father came directly from his mother, who had been a schoolteacher in Pratt, Kansas, before she married O. W. Little gems like "There is no R in wash," "People who don't listen have to feel," and "You've made your bed; now you have to lie in it," were grammatically correct expressions that Pop repeated to all of us Bonner children when we were growing up.

Eager to drive Mom to the fossil beds for one of their early dates, Dad knew spending time alone with her would win her over. As the two traversed the Wichita County flatlands to the Logan County chalk bluffs, the transformation was almost magical. The farmland around Leoti is fertile and level, the surrounding view expansive and unending. North of Leoti, however, the surface is more undulating, a wavy plain that resembles the ocean. Early explorers and visitors, traveling by prairie schooners, were struck by the vastness of this land. Like them, Dad saw the rolling landscape as a vast ocean. As my parents drove to the outcroppings, Dad delighted in describing the irony that the Cretaceous inland sea that had existed millions of years ago had gradually changed into the "prairie ocean" they were now traversing.

Miles away from any other people, Dad pulled off the road and maneuvered slowly across a pasture. The wide-open space seemed incredibly quiet at first, but Mom realized that if she listened carefully, she could hear flurries of activity, starting with the *shush, shush* of the wind through the tall grama grass. Dad helped her out of the car, and when she stood to her full height of five-nine, her brown eyes gazed into his blue ones. They walked toward a smooth knoll that was perfect for their picnic. Hiking toward the cream-colored bluffs, she heard the raspy, strumming sound of insects—buzzing horseflies and snapping grasshoppers. She felt attuned to nature and sat quietly on the knoll, where she took sandwiches out of the lunch basket she had prepared.

"See those beds over there?" he asked, sitting down beside her. "The chalk making up those ravines and bluffs was deposited here millions of years ago." Mom listened, fascinated, as he spouted off the major eras of geologic time he had learned from Mr. Wedel. The eras of geologic time measure the full extent of the Earth's existence: Precambrian, Paleozoic, Mesozoic, and Cenozoic.[3] Dinosaurs, giant marine reptiles, and the earliest birds appeared in the Mesozoic Era, he told her.

After lunch, they set off to scan the bluffs for fossils. "You can find lots of vertebrae, fins, and shark's teeth, but it is rare to find a complete fish," he explained as they walked together down a gully. "Imagine all the forces of nature preventing these fish from being complete. First of all, there were predators like mosasaurs, tearing them apart for dinner, which is how most of them died. Then would come scavengers like sharks, snacking on the remaining parts of the fish," he said, picking up a shark's tooth.

The strong wind tugged at the scarf tied over Mom's black hair, and she shuddered. She realized her new boyfriend definitely had a way of making things sound intriguing. "Finally, there was the sea itself," he continued. "The waves slowly worked away at remains, the silty deposits coming in with each wave. You can think of this as a fossil graveyard." Mom noticed that the area was littered with thin, dark fish scales. Dad rarely picked up the fragile scales—to him they were not significant because they were not connected to a larger fossil. In later years, some of the paleontologists Dad met were extremely interested in studying the scales. For now, Mom picked one up here and there but left most of them where they lay.

The couple continued walking around and talking, him giving her a hand when she needed to step over gaps in the ground. When they walked up a shallow hill, he accidentally slipped a little and they collided, laughing. He showed her the rock hammer he used to break up "overburden" material when he had found a specimen. It looked like a geologist's hammer for breaking rocks but was actually a bricklayer's tool designed to break bricks in half.

Dad spotted a giant shark's tooth and excitedly picked it up. "Speaking of sharks, look at this beaut!" he said. The two sweethearts scoured the ground around the two-inch-long tooth to be sure it was simply a single, random tooth and that there was no evidence of more shark material either eroded along the gully or coming out of the bluff above.

"Those other shark teeth we found, the small serrated ones, are from *Squalicorax*," he told her. "This one is an entirely different species. It's an *Isurus* [now called *Cretoxyrhina*] shark." He handed it to her. "Here, you keep this. A beautiful tooth for a beautiful girl." He was so caught up in that moment that he left the rock hammer on the ground. He didn't think about the tool until several days later. The next time he went to that same area of the badlands, he looked high and low for it, but it was gone.

Dad further explained to Mom that the shark he called Isurus, now *Cretoxyrhina* ("Cretaceous sharp nose"), was a sizeable cartilaginous fish that resembled a modern Great White shark. An apex predator with large smooth teeth, it was one of the biggest sharks of the time, attaining lengths of twenty-six feet. It usually fed on mosasaurs, plesiosaurs, sharks, and other fish. By comparison, the smaller *Squalicorax* rarely reached the length of ten feet.

In addition to their natural attraction to each other, it was the fossil beds that sealed the deal. Marion and Margaret dated for a few months, then became so serious that they started talking about getting married. She was a devout Presbyterian who attended church every Sunday. He had been raised a Methodist, but dabbling in paleontology had led him to be agnostic. He told her he wouldn't be going to church with her, and she accepted that. She told him she wanted to raise their children in the Presbyterian church, and he accepted that. One day in September 1935, they traveled west, to Greeley County, and were married by the justice of the peace in Tribune, Kansas. Mom was eighteen, and Dad was twenty-four. Clare Berg stood up for them as their witness. The sweet, smart girl and the enthralling flirt would forge a stimulating life together.

There is no wedding photograph of my parents. Instead, we have multiple photos that were taken around that time, and of the children coming along every few years.

The Stock Market Crash occurred in 1929, and the Great Depression followed. The Depression lasted a full decade, from 1929 to 1939. Dad's fossil hunting was severely curtailed during those years because he spent his waking hours scrounging for work. He didn't get back to his favorite pastime until the end of the 1930s, when he found and collected a complete *Albula* (a small fish now called *Apsopelix*).

Pop continued to help in his parents' two ventures, the theater and the farm, but he added other options, including hunting and trapping, house painting, sign painting, and working as a fry cook. In their early marriage, Mom and Dad lived briefly with Clare in her small square stucco house on Fifth Street, but they soon found a little rental house on Main Street (Kansas Highway 25) in the block south of the Wichita County Courthouse.

Our Mom and Dad's courtship and early marriage took place during the height of the Dust Bowl as well as the Depression. What a shock it must have been for the city-dwelling Bergs to have immigrated to Wichita County during the Dirty Thirties. Leoti was in the midst of the largest man-made natural disaster in the nation's history.[4] Soil scientists estimated that during the storms of March and April 1935, four to seven tons of dust per acre fell on western Kansas.

Sod-busting, when combined with severe drought, led inevitably to the Dust Bowl. Nature had covered the High Plains with a dense sod—buffalo grass—that withstood drought and supported vast herds of bison. While this protection of the topsoil had existed for thousands of years, it took only a few decades for immigrant farmers like the Bonners to turn over the soil for crops and inadvertently set the stage for catastrophe. By 1933, the interior grasslands of the Continental

U.S., some two hundred million acres, had become wheatlands. High commodity prices and high demand created a wheat boom. But the horrible windstorms of the 1930s, combined with severe droughts, blew away the soil.

Older paleontologists noted the effect of the western Kansas wind. In the 1930s, Barnum Brown, a renowned dinosaur collector who had grown up in Kansas but was now at the American Museum of Natural History, read about the disaster occurring on the plains from his office in New York. He said that the blowing dust would uncover countless fossils.[5] In the softer parts of the chalk and shale, the wind and water erosion did expose many new discoveries. It should be noted, though, that fossil exposure through erosion is more evident in the broken-up areas of exposed sediments, such as along gullies. Areas composed of hard chalk or protected by caprock take a very long time to erode. That is why Monument Rocks in Gove County, the well-known Kansas "Chalk Pyramids," look much the same today as they did one hundred years ago.

Barnum Brown (1873–1963) was from Carbondale, a small town south of Topeka. He was educated at the University of Kansas and introduced to paleontology by renowned KU scientist Samuel W. Williston. During his college years, Brown worked in South Dakota and Wyoming, where he found a skull of the Cretaceous horned dinosaur, *Triceratops*. He spent most of his career at the American Museum, and in 1902, leading a trip to eastern Montana, discovered the first skeleton of the gigantic carnivore, *Tyrannosaurus rex*.[6] Brown was a "character" among paleontologists. Sporting a full-length fur coat on fossil hunting expeditions, he became known as "Mr. Bones." Much later, at one of Dad's more thrilling Society of Vertebrate Paleontology meetings in New York, he met and shook hands with Brown.

European Americans were lured onto the Great Plains by multiple popular theories and inducements. In addition to myths of "rain following the plow" and societal encouragement to leave the teeming cities (such as in Horace Greeley's famous exhortation, "Go West, young man!"), doctors often suggested western air as a cure for lung and heart diseases. An early plains traveler, Josiah Gregg, was one such patient who headed west in 1831 on the advice of his physician to help ease his consumption (tuberculosis).[7] This relocation therapy became popular, and "thousands of Americans seeking a cure (often a last-ditch cure) for their tuberculosis migrated west from the 1840s to the 1920s, persuaded by physicians who touted the region as a curative Eden."[8]

Margaret Berg had rheumatic fever as a child. As an adult, Mom rarely talked about it, but in later years Pop told me that after she recovered, her Kansas City doctors speculated that she might not live past the age of eighteen. It is now known that this condition can spring from strep throat and its treatment is strong antibiotics, but in the 1920s, rheumatic fever was the leading cause of death in children in the United States. One of the serious complications of rheumatic fever is long-term heart damage, or rheumatic heart disease. This disease weakens the valves between the chambers of the heart, and severe cases can lead to death. At the time of Mom's diagnosis, however, medical science could not yet predict heart disease from rheumatic fever, and current diagnostic tools such as EKGs were not yet invented. (Mom also suffered from asthma and allergies, which she passed on to many of her children.)

What is clear is that our mother easily outlived her majority. Grandmother Berg had brought her family to Leoti to take care of her parents, but perhaps she also considered Mom's condition. Doc Doyle may have suggested that the air of the West could alleviate some of Margaret's ailments (not aware of the supreme irony that the Dust Bowl was just around the corner).

Despite her childhood illness and its grim prognosis, our mother lived to the age of fifty and had eight children with the Old Man of the Fossil Beds. She enjoyed the outdoors to the fullest and became an avid gardener, fossil hunter, and nature lover. When Mom and Dad were able to take day trips to the fossil beds in the 1930s, they found them to be a respite from the dust storms that were appearing with alarming regularity. The lands around the Smoky Hill River were unbroken, the grasses holding down the dirt. Here they could breathe a little more freely.

Dust Bowl Survival

1930s

THE TERM "DUST BOWL" aptly applied to what remained of the south central United States after a prolonged drought and unrelenting wind carried away vast volumes of its rich topsoil. According to later estimates, more than 1 billion tons of topsoil were lost across the Great Plains, primarily between 1934 and 1935, the most severe drought period of the event. In *Great Plains,* author Ian Frazier describes the horrific conditions of a major dust storm that occurred in 1934, the year Mom graduated from high school and that she and Dad began dating:

> In mid-April, a giant dust cloud, black at the base and tan at the top, rose from the fields of eastern Colorado and western Kansas and began to move south. Inside the cloud, darkness was total, and remained for hours after the cloud passed. People in the cloud's path thought the end of the world had come, and went to churches to await it. The storm left dead birds and rabbits in its wake, and drifts of dust six feet deep against the sides of houses.[1]

At its peak, the Dust Bowl covered 100 million acres of the semi-arid center of the country.[2] Much has been written about farmers who left the land, ruined by the collapse of the economy. But according to writer Timothy Egan, "Not much was heard about the people who stayed behind. ... [M]ost people living in the center of the Dust Bowl, about two-thirds of the [affected] population in 1930, never left during that hard decade."[3]

The editor of the *Leoti Standard* described, in a story called "Black Monday," an eerie dust storm that occurred in Leoti in 1933:

> [B]y 3:00 p.m. lights were turned on everywhere. By 3:30 the day was dark as an ordinary night, and then for about 45 minutes, it seemed that an additional black pall was cast over everything,

the sky becoming velvety black. ... Dust was being kicked up by the local wind also. About four o'clock, a handful of hail with a little more rain fell here in Leoti. The hail, by the way, was gray, dust being thoroughly mixed in the ice. ... We have found no one who ever witnessed a similar occurrence, here or elsewhere. The darkness was equal to that of a total eclipse. Old timers couldn't recall a parallel case.

Once the rest of the nation realized the middle of the country was being blown away, pressure built for something to be done about it. After grappling with many possibilities, President Franklin D. Roosevelt instituted federal programs to reestablish the topsoil. These included the New Deal programs of the Work Progress Administration (WPA) and the Civilian Conservation Corps (CCC), whose workers planted windbreaks between sections of land to try to tame the force of the wind and hold down the soil. To this day, still visible around many of the farms of western Kansas, are shelterbelts from the Dust Bowl era, consisting of such drought-tolerant trees as cedars, hackberries, and cottonwoods. The Soil Conservation Service, created in 1935, encouraged better agricultural practices, and Soil Conservation districts paid farmers to let some of their acreage return to grassland.

Grandmother Berg saved this postcard from the Dirty Thirties in her photo album.

The New Deal programs were a lifesaver for some families—keeping them going by giving them a subsistence wage. In putting people to work during the Depression, the programs provided a measure of self-respect over simple cash relief. The National Youth Association (NYA) was a program for young people from ages sixteen to twenty-five to earn a little money. It operated from 1935 through 1939 under the WPA.[4]

Dad volunteered to serve as Wichita County's director of the NYA in 1936, and some of the structures he and his "NYA boys" built were in the City Park on the southwest side of Leoti. Citing the record books of the Civic League in Leoti, the local paper reported: "Time rolls along and ... there was a period when the park received little attention. Then, under a government youth movement, under the direction of Marion Bonner, the shelter house was erected and the stone fences built." The City Park was only a few blocks west of the theater, and later the Bonner house in Leoti, so we kids spent countless hours at the park playing on the steel slides, swings, teeter-totters, and merry-go-round. Dad's service with the NYA opened his eyes to public service. He would later take a volunteer position during the World War II years as chairman of the bond drive.

The largest structure in Leoti built during the New Deal period was a limestone municipal auditorium, city hall, and fire station, which was constructed by Wichita County's WPA in 1940. The building now houses the Museum of the Great Plains and the Wichita County Genealogical Society. It was built in a "federal or institutional style" and is distinctive because of its material of construction: Logan County limestone. It was the last and also the largest structure built using Logan County limestone.[5]

The most severe waves of the Dust Bowl in western Kansas, eastern Colorado, and the panhandle of Oklahoma occurred in 1933 and 1934, then again in 1936, around the time of my oldest brother's birth. Another harsh storm followed in 1939 before the dust of the High Plains finally, literally, settled when the New Deal programs started taking effect and the rains came at last. Dad told us in later years about times when he and Mom were out riding around in his Model T (one that he bought after liking Mr. Wedel's), and he would reach over and gently wipe away the dust that had accumulated on the flares of her nose.

Shortly after their marriage in 1935, they moved into their small rental home on Main Street. Its framed wood windows and doors were drafty when the robust winds started up, but in that era, no house could have kept out the dirt that sifted in from the storms. Expecting their first child, Mom was concerned about keeping the dust out of their home. She placed wet cloths around the windows and, after "Orvie" was born, draped the infant's bed with a wet sheet that picked up a layer of dust overnight. In the mornings, she soaked the muddy cloths, rinsed them thoroughly, and put them in place for the next round of winds. Generally, the mornings were the clearest part of the day, but the inexorable wind started blowing in the afternoons.

A bizarre byproduct of the environmental disaster was the explosion of jackrabbits. They were one of the few animals that thrived during the Dust Bowl's warm weather and lack of rainfall. Matching the worst years of the Dust Bowl, the worst years of the rabbit invasion were 1934 and 1935, when they ate up every plant in their path down to the roots. Farmers were desperately fighting to save their crops, so communities in western Kansas held "rabbit drives" on Sunday afternoons in which they drove the rabbits into an enclosure and clubbed them to death. It seems a gruesome activity, but it was necessary to control the overpopulation.

Dad trapped badgers and raccoons for their hides to supplement his income. Raccoon hides fetched a dollar and a half apiece. He said he sold these hides to pay for Orville's birth. The little yellow house south of the library is Orv's first memory of life in Leoti. Born in April 1936, he was named after our paternal grandfather, Orville Wesley Bonner. The family and townspeople called the elder Bonner O. W. and the little namesake Orv, or Orvie.

Mom, seven months pregnant, was honing her cooking skills in their small kitchen while Dad and Grandmother Berg sat in the living room relaxing before dinner. Dad had spent the day painting houses and had just cleaned himself up outside at the water pump. He was exhausted.

"People are really struggling, Clare. I'm getting a cut of the profits from the picture show, but I feel like I'm doing all the work there. The theater is supporting my parents, sisters, and brothers. I've been painting, trapping, and sometimes working as a fry cook, but it seems like it's not going to be enough."

"Marion, you are doing very well. I wouldn't worry about it if I were you," Clare responded, realizing that at this moment he needed someone to listen, more than give advice.

"I'm learning that there's a difference between eking out a living as opposed to earning money doing something I truly enjoy," he said. "I wish I could be like the Sternbergs, but the sale of fossils is just not lucrative. 'Fossiling' is what I like to do more than anything, but it won't feed us." He sighed. "Anyway, I shouldn't complain. It seems like nowadays everyone is living hand to mouth."

"You're right. These are tough times, but this won't last forever. You will be able to find work you enjoy," Clare replied. "Things will change." In her mid-fifties, Clare had the wisdom of surviving many of life's setbacks.

"Have I ever told you what a wonderful woman you are?" Dad asked, winking. "If you were younger, I'd have had my eye on you!" The two of them laughed at this standard repartee.

Winning Clare over had been extremely important before he and Mom married, and he was glad his mother-in-law was so affable. For her part, Clare was fond of Marion's dramatic, roguish nature. She realized that paleontology was his true calling and that he would return to the fossil beds whenever he could. She also knew he would find ways to earn money in the small town of Leoti. None of them had the option to leave.

"We will make it," a voice called from the kitchen. "Dinner's ready!"

The wind and dust helped scour fossil bed gullies for new discoveries, but in addition to fossils, Dad and Mom started finding arrowheads and other artifacts the Indigenous plains cultures left behind. They found spear points of the oldest known Great Plains dwellers, the Clovis people, who were active between 11,050 and 10,800 years ago. Clovis hunters used a distinctive type of point to kill mammoths. These spear points were four to five inches long, two inches wide, and one-third inch thick, with an extremely sharp edge. After the mammoths were gone, another culture emerged, using the Folsom point to kill prehistoric bison. The Folsom people, active from 10,800 to 10,200 B.C.E., used spear points that were shorter than Clovis points but had longer flutes and different flake patterns. More recent arrowheads, knives, and scrapers in their collection are harder to classify but likely date from the Early Plains Archaic, a pre-European period.[6] My parents found examples of artifacts from all three of these Paleo-Indian cultures, plus more recent ones from the Plains Indian tribes.

Even if the U.S. government and the influx of white settlers had not pushed out the Indigenous people, the decimation of the Plains Indians' means of sustenance did much to reduce their numbers. As historian Kenneth S. Davis wrote, the natural environment "had been so transformed by the white man in western Kansas that the Indian could no longer have lived there, as he had in the past, even if the white man had abruptly withdrawn. For by 1878 the buffalo, upon whose abundance the plains Indians' life and culture were wholly dependent, had disappeared from Kansas and was rapidly approaching total extinction."[7]

Many of the chipped stone items in Mom and Dad's collection came from nearby counties, but Wichita County alone had a half dozen known archeological sites.[8] Their amateur collection represented but a small sample of Plains Indian activity in western Kansas.

Two years after Orv's birth, Mom and Dad had their second child, my oldest sister Clare Jane, born in February 1938. The little daughter was named after our maternal grandmother, Clare, and her middle name came from an aunt, Ida Jane Berg. Because there was a well-established Clare in the community, Mom and Dad called her Clare Jane, which stuck for life.

A year after Clare Jane was born, the last severe dust storm rolled through Wichita County. The nationwide drought of the 1930s was finally ending. By the time of the last storms, my parents had become old hands at coping with "blow dirt." Like other plains dwellers, they celebrated when the rains came. The return of sunflowers was a happy sight, and the sound of meadowlarks lifted their spirits. Like dust storms, Kansas thunderstorms can be seen for miles and miles. Their approach brought joy rather than dread.

The sunflower, the state flower of Kansas, is abundant on the plains, but more so along farm fences. It springs up where the sod has been broken up for farming. These are not the seed-bearing variety of sunflowers but the smaller wildflower variety. When Pop met people who weren't from Kansas, he liked to quote an old Wichita County farmer, who, upset with a transaction, said: "Kansas is known for three things—sunshine, sunflowers, and sons-of-bitches. The latter you will find in the back end of the Co-op."[9]

Our encouraging and talented mother was a big reason our family was able to cope. She was a nurturer who taught all her children crafts and skills, including sewing, painting, and writing. After arriving in Leoti, she started using the materials on hand. Dad said she was smart as a whip and able to "make a silk purse out of a sow's ear." From the fossil beds, she collected natural materials for dried arrangements and was an early proponent of recycling and reusing whatever she could.

Mom learned how to sew at Grandmother Berg's elbow, initially using a treadle sewing machine. She made outfits for all of my brothers and sisters and saved scraps, turning them into clothes, quilts, anything useable. The ability to make costumes for plays, decorations for parade floats, and items of home décor was much in demand after Mom joined the Mothers Study Club, which started in the forties before World War II. During the war years, she was active in Women and Children First (WCF). These clubs were an effective way for local women to socialize, gather together to perform community service, and provide advice and comfort for each other. The pooling of remnants was the basis for many club-created quilts as well.

Once the children started coming, Dad crooned to the babies versions of songs he had heard as a child. He also liked to recite rhymes that ended in tickles, such as one he repeated when gently touching the forehead, eye, nose, and chin of a child: "Knock on the door (rap on forehead), Taaaake a peek in (lift eyelid), Lift up the latch (pull up nose), and walk right in! (tickle under chin)."

He also sang tunes from the movies and poems from sources like *Captain Billy's Whiz Bang* (his generation's version of *MAD* magazine):

> Look at those eyes, look at that nose, look at those baby feet
> Walkin' down the street, Hey, fellas, ain't she sweet?
> I wonder who that baby is? Look out! She's coming near!
> Yoo hoo, Woo hoo—why, imagine that!
> Well, hello there, Grandma dear!
> (*lowered voice*) ... Boys, I'd like you to meet my Grandma.

While the dinosaurs of the Mesozoic Era (the Triassic, Jurassic, and Cretaceous periods) have captured the popular imagination for more than a hundred years, the "sea monsters" of Kansas are equally fascinating, although not as well known. The giant swimming reptiles in the mosasaur and plesiosaur families were themselves impressive and held a special place in the fauna of the Western Interior Seaway. In the *Jurassic Park* fantasy movies, viewers meet all kinds of prehistoric beasts, but it is not until *Jurassic World* (2015) that we see a mosasaur leap out of its tank to consume a *Pteranodon*. Then, in the opening scene of *Jurassic World Dominion* (2022), a massive mosasaur blasts up from under the sea to attack a fishing vessel. And in 2025's *Jurassic World Rebirth,* a massive mosasaur is one of the show's main stars.

The mosasaurs were relatives of sea snakes and lizards. Their gigantic jaws were double-hinged like those of snakes, and their conical teeth were like those of modern crocodiles. The largest modern crocodile has been measured at twenty-three feet long, whereas the largest mosasaurs reached fifty-five feet long. The three main genuses of these successful predators in the Kansas seaway—*Tylosaurus, Platecarpus,* and *Clidastes*—were distinctly different from the mosasaurs found in Europe. The Kansas mosasaurs appeared much earlier in the lineage.[10] European mosasaurs, found first, actually occurred later in geologic time, in the late

Cretaceous. The name mosasaur came from the Latin *Mosa* for the Meuse River, and Greek *sauros*, or "lizard."

The largest *Tylosaurus* that had been found by the 1930s was discovered by Charles D. Bunker, a University of Kansas naturalist, in Wallace County, in 1911, the year of Dad's birth. Looking at the size and amount of fossil material, Bunker originally thought it was a plesiosaur, and the specimen was not completely collected until Handel T. Martin's party of paleontologists went back and got all the bones.[11] Today a mounted cast of the large reptile, known as the "Bunker Tylosaur," hangs over the entry rotunda of the University of Kansas Museum of Natural History in Lawrence. At forty-five feet long and mounted in a long, snakelike curl, the awe-inspiring fossil looks out upon viewers with a gaping leer. The extinct reptile has become legendary: "In 2014, more than one hundred years after [paleontologists] returned to the University of Kansas bearing wooden crates of the fossilized bones of the sea serpent, the governor of Kansas signed into law a resolution making *Tylosaurus* the state's official marine fossil."[12]

Michael Everhart described the creature in his book, *Oceans of Kansas:* "*Tylosaurus prorigor* was the largest of the mosasaurs in the Western Interior Sea during the deposition of the Smoky Hill Chalk, reaching ten meters" in length. "The larger mosasaurs, such as *Tylosaurus*, preyed on a variety of marine vertebrates, including fishes, plesiosaurs, birds, and even other smaller mosasaurs "[13] Pop found skulls and portions of *Tylosaurus* in the early years, but it wasn't until 1979 that he discovered a nearly complete specimen.

Another marvelous beast of the Cretaceous Sea was the long-necked plesiosaur, *Elasmosaurus* (the mythical Loch Ness monster is suggested to be a late surviving long-necked plesiosaur). Dad always wanted to find a significant fossil of this genus. Over the years, he happened upon a scattered bone here and there, but never found the larger specimen going back into the rock. In the Smoky Hill Chalk, this sea monster was rare, but in the adjoining Pierre Shale, Dr. Theophilus Turner found a nearly complete skeleton in 1867 near Fort Wallace. This was the type specimen of *Elasmosaurus platyurus,* described by E.D. Cope in 1868.[14]

In the years to come, Pop and my brothers had much more success finding significant specimens of the Cretaceous short-necked plesiosaurs, *Dolichorhynchops* and *Polycotylus.*

Fossil tools were an important part of the trade. Initially, for close work in the chalk and shale, Dad used tools such as jackknives and modified screwdrivers. The smallest awls and scrapers in his toolkit were good for detail work on the fossil after it was collected. To dig out a quarry or remove the overburden before collecting, he used shovels, large picks that resembled miners' picks, and geologists' rock hammers. He had moved past that first bricklayer's tool that he lost while hunting with Mom. Surprisingly, in 1970, almost four decades later, he found that first tool. It was deeply weathered but still in one piece. My brother Chuck donated it to the Wichita County Museum in Leoti, where it is in a display case labeled, "Marion's Rock Hammer."

When Pop became good friends with George Sternberg in the 1950s, the Fort Hays State University scientist gave him a Marsh pick, which is the entrenching tool shown in the Society of Vertebrate Paleontology logo.[15] The Marsh pick handle was twenty-two inches long, and Dad often included the pick in photos of fossils for a size reference.

Other paleontologists invented similar tools. Barnum Brown modified a miner's pick that he called a "dinosaur pick." Sternberg also loaned Dad some small scraping tools, one of which he called a Broili tool, also named for a paleontologist.[16] In a dated December 17, 1955, Sternberg wrote: "By the way, Bonner, I have a very fine small pick for you, so the next time you are over [to Hays] will you bring the one you have over and I will give it back to Rouse and give you this one for keeps."[17]

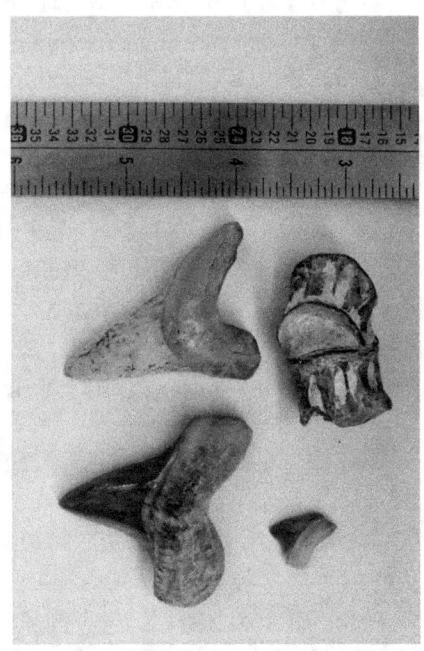

Random fossil finds: two Cretoxyrhina shark teeth (left); two Ichthyodectes vertebrae and a Squalicorax tooth (right).

Remembering George Sternberg's penchant for teaching youngsters, in the 1930s, Pop encouraged Leoti High School students to take an interest in paleontology. The *Leoti Standard,* in a story called "Fossil Hunt," dated April 23, 1931, reported:

> Some time ago, Marion Bonner conducted a fossil hunting trip to the hills of Logan County, taking two representatives from the science class, Edwin Rutt and Wayne Hayden, with him. They

discovered a fish tail or caudal fin of a fish of the genus *Empo*.

Plaster of Paris was poured over the specimen while in the rock and then allowed to set into a hard mass. After it had set, the rock was dug out from under the specimen and the slab of plaster was turned over; the specimen was set perfectly in the cast. It has now been mounted and is on exhibition in the science room.

On one similar hunt, however, a mosasaur disappeared. On this occasion, right before Dad went into the Army, he took with him a Leoti lad who was excited about hunting fossils. Happy that another high school student was curious about paleontology, Dad took the young man to the massive outcroppings Samuel W. Williston called Castle City (which is now New Jerusalem Badlands State Park). This site had many canyons of chalk to explore, and that day, Dad found a *Platecarpus* mosasaur. He dug it out to see how complete it was. It wasn't a full skeleton but had some skull material, several feet of the vertebral column, and a few ribs. Dad covered it with flat pieces of chalk and told the boy he would come back another day to collect it.

For the young friend, the lure of the sea creature was too much. He decided to return to the site with several of his schoolmates the next day. They dug out the bones without Dad's guidance, and when Dad returned a few weeks later, he saw that most of the fossil was gone except for a few pieces that had broken while the boys excavated it. The high-schoolers took the mosasaur bones to the local newspaper editor, and they laid near the front window of the *Leoti Standard* office on Main Street for years. Dad never said a word to the young man about it, but he didn't take him fossil hunting again either.

Privately, he felt sick about the loss of this mosasaur. He had hoped to collect the entire fossil in burlap and plaster jackets and send it to a museum. At the time, the taking of the fossil seemed tragic, but Dad later realized that mosasaurs were fairly common. What the episode did for him was to create a sense of privacy and caution. He decided not to trust a dabbler ever again and developed a sense of protectiveness for the rest of his career in the fossil beds. This scientific pursuit was serious for him, no longer a hobby, but a vocation. And a seed had been planted that would grow into the Old Man of the Fossil Beds stewardship.

Dad continued to help support his parents and extended family during the Dirty Thirties. He purchased a sturdy family sedan, a Studebaker Commander, from his brother-in-law, Ralph Berg. While O. W.'s sale of the family's farmland had funded the purchase of the movie theater in 1926, during the 1930s, the patriarch stepped back from an active role. The Bonner family's primary business became the motion picture theater, and increasingly Mom and Dad pitched in, Dad running the projector and Mom selling tickets and popcorn.

Throughout the Dust Bowl and the Depression, people who could afford it escaped to the theater. With adult admission at twenty cents and children a dime, the show offered western Kansas folks an affordable break. Movies provided a social outlet for the town and also came to define the Bonners of Leoti, first the family of O. W. and Viletta, and later that of Marion and Margaret. A steady business through the Depression meant the Bonners had the means to move to a larger building. In 1937, O. W. purchased a long building on the corner of Main and L Streets. He and his children would remodel this building, the former Leoti Auditorium, and rename it the Plaza Theater. As the theater expanded, so did Mom and Dad's family. Their brood continued to grow with the addition of three more children in the 1940s.

Nineteen-thirty-nine would prove to be a huge year for bookings of Technicolor movies from Warner Brothers. First, it was the year of that quintessential Kansas movie, *The Wizard of Oz*, based on L. Frank Baum's novel. Also in that year, my parents met a handful of marquee stars in person: Errol Flynn, Olivia de Havilland, and Alan Hale, Sr. at a premiere in Dodge City for the movie *Dodge City*. Later that year, they went to a premiere of *Gone with the Wind* in Garden City and met Hattie McDaniel, Olivia de Havilland (again), and Clark Gable.

Hattie McDaniel, from Wichita, Kansas, would become the first Black woman to win an Oscar for her role in *Gone with the Wind*, despite the racist typecast figure she played of Mammy. Because *Gone with the Wind* was so popular in western Kansas, Dad booked it a total of seventeen times, and people from many surrounding counties drove to see it. To modern sensibilities, the film's stereotypes are jarring and it has been criticized for its glorification of the Old South and slavery.

Dad saw "GWTW" as a saga of an era long passed, and being from Yankee stock, did not view the end of the Old South as a tragedy. He greatly enjoyed the scene when Rhett Butler walks away from Southern belle Scarlett O'Hara with the famous line, "Frankly, my dear, I don't give a damn." Beyond the movie's storyline, however, Dad identified with the metaphor of the wind. One of the title cards of the movie read, "And the wind swept through Georgia. ... " Another one summarized, "A civilization ... gone with the wind."

The wind was ever-present in western Kansas and always brought change. After Mom's death in 1967, Pop rhapsodized about how, when the two of them were both "gone with the wind," that that phrase would be a fitting epitaph for them.

PART II - FAMILY

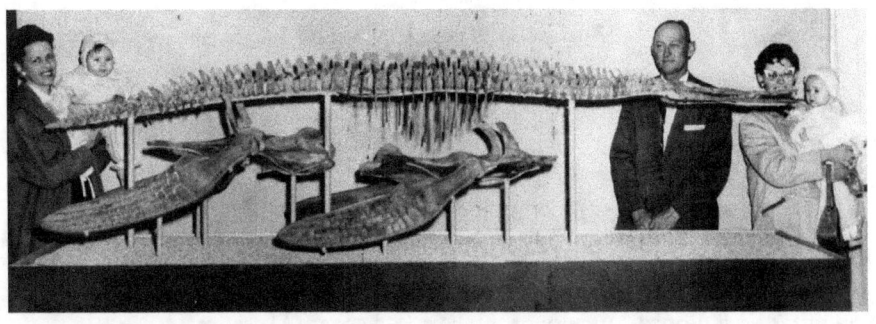

Previous page: The nearly complete *Dolichorynchops* plesiosaur discovered by Marion Bonner in 1955 and donated to the Fort Hays State University museum. Flanking the fossil, left to right: Margaret, Melanie, Marion, Clare Jane, her daughter Kari. Photo courtesy of the Fort Hays State Museum.

Fish Tales
1970s

IN THE 1970S, AFTER Mom passed away and my brothers Chuck and Dana had both gone to college in Hays, Dad and I had my four high school years together. During this time, I knew he was struggling. Money was scarce, but he was holding on, grinding toward the finish line of his child-rearing marathon. Every now and then he sold a nice fossil for a pittance, but he stubbornly refused to work for anyone. All of us Bonner kids developed our own forms of stubbornness. When one of us did something contrary to his suggestions, he'd mutter, "Independent as a hog on ice." He knew whereof he spoke.

He drank a lot more during these years to cope with the judgment he felt from the small town's citizens. Having been a self-made man for so long and a successful movie theater owner, he didn't want to work for anyone else. I was defensive of his choice when well-meaning folks voiced their concern. He didn't want to be beholden to anyone. Pride wouldn't allow it. I understood.

Because I grasped Pop's point of view so well, this understanding also allowed me to tease him as much as he teased others. I met him quip for quip and doled out a fair share of smarty-pants comments (which stunned my older siblings), but he always went along with it, laughing with me. He didn't have to pretend to be something he wasn't around me.

Despite his ego, he was caring and consistent, even though we were poor. We were still able to eat, but it was a simple diet. He made me lunch for school every day, usually a peanut butter and jelly sandwich with a jar of sweet tea and a boiled egg (adding a small foil pouch of salt and pepper for the egg). Most of our fruits and vegetables were canned. He called me his "pancake girl" because we had pancakes for breakfast every weekend and some weekdays. These were not fluffy flapjacks; he used thin batter and they came out the size of a plate. When he ceremoniously carried them from the griddle to the table on a wide spatula, he'd drop them on my plate and yell, "Horse blankets!" In consistency, they were more like large crepes than traditional griddle cakes.

Accompanying the pancakes was camp coffee, made by bringing grounds in his black-crusted aluminum pot to a boil, letting them settle for a good five minutes, and then carefully pouring the thick liquid into our mugs. It was almost as strong as espresso, but we added lots of milk and sugar. I started drinking Pop's coffee in fifth grade and that beverage got me through life until 2012, when a peptic ulcer screamed loud enough to make me quit coffee, cold turkey. Even so, the smell of strong coffee will always be one of my favorites. It reminds me of my childhood and my caring father. Evidently it didn't affect my growth because by the time I graduated, I was the second tallest in my class, one inch short of my adult height of five-eleven. People called me "Stretch" and "Long, tall drink of water." I was taller than Orv, Dad, and my sisters.

Pop and I at a family picnic in Arizona, 1973. Photo by Bud Bristow.

After I left Leoti in 1975 for the University of Kansas in Lawrence, Pop began hearing news about the fossils he had sent to Los Angeles the decade before. Paleontologists at the Los Angeles County Museum[1] were working away on the Bonner fossils, and many of the specimens were being prepared for public viewing. On January 10, 1976, Dad's friend Graham Berry, a California-based writer, sent him a letter about a new exhibit at the museum. Berry described the installation, a wall of fossils, in detail because he was not sure Dad would be able to come to California to see it in person:

> The exhibit extends along an entire wall of the La Brea Room and has a *Pteranodon*, turtle, shark, *Gillicus arcuatus,* primitive bony fish, *Xiphactinus audax,* and a 12 ½-foot mosasaur made of parts found by you and George Sternberg. ... The entire exhibit has a "headline" over it in large letters reading: "Prehistoric marine animals of a fossil sea—the Bonner Collection." ... The visit to the museum was almost like having a visit with you. Almost, but minus your wit. ... The exhibit, I understand, will remain in the La Brea Room until the museum completes a large Mesozoic Room, wherein many of your fossils will be exhibited on a permanent basis.

The museum kept the Bonner Collection up for more than a decade, from 1976 through 1987. Unfortunately, Berry was right. Dad was never able to make the trip to the West Coast to see it, but other friends and family sent him photos, and the *Los Angeles Times* did a story on the exhibit. Closer to home, Kansas newspapers covered it as well. "Leoti Man's Fossil Finds on Display in Los Angeles" read the headline of a story in the *Salina Journal*,[2] and the *Hays Daily News*'s story read, "Californians Inspect Western Kansas Fossils."[3] The *Salina Journal* story quoted the museum's brochure, which focused on Dad's dual vocations: fossil hunting and the theater: "The fossils now on exhibit are a part of the many collected by Marion Bonner. ... A successful movie house operator by night, Bonner had free time during the day and a growing family to help in the search. ... Today, Mr. Bonner is honored as one of the great professional fossil hunters. He says he owes it all to William S. Hart." Hart was a cowboy star of the silent era.

The fossil wall was still up when fellow Kansan J. D. Stewart was hired as a curator for the Los Angeles museum in 1986. Stewart became a pivotal scientist putting forth theories about Dad's discoveries and, crucially, rediscovering fossils that were in storage in the museum.

In 1976, with his wife and children gone, Dad decided the house in Leoti was too big for him. He was ready to leave the town where he had lived since 1919, and he strongly desired to be closer to the fossil beds. He put the Leoti house on the market and waited to sell it before buying a small home in Healy, an unincorporated town in northwest Lane County (in the 2020 census, Healy and its surrounding area reported 195 residents). The Bonner kids who could make it to Leoti in the fall of that year helped him move to his house in Healy, which he bought for the low price of five thousand dollars. This house was twenty miles closer to where my brother Chuck and his wife, Barbara Shelton, lived.

No one had been living in the Healy home, located north of Kansas Highway 4 on a curb-less dirt road. It was a two-story stucco structure with a steep gabled roof that looked like it had been built in the 1920s—sturdy and primitive. We were surprised to see, upon moving Pop to the new place, that the previous owner was storing seed corn in the living room. Before long, with the help of his children, the house was habitable. Clare Jane made a visit back to Healy in the summer, bringing curtains, paint, and wallpaper, and we pitched in to decorate the house to make it look more livable.

A few years later, in 1981, Dad bought a twenty-seven-acre tract of land at the even tinier town of Shields on Highway 4 in Lane County, nine miles east of Healy. The house was uninhabitable due to termites, but the Quonset hut on the Shields property gave him an expanded place for preparation and fossil storage. Serving as his last laboratory, the big shed had windows and electricity, so he could see well during preparatory work. In his later years, the space was the site of prep work on exceptionally large fossils: *Xiphactinus, Polycotylus,* and *Uintacrinus. Uintacrinus* (crinoids) are fascinating invertebrates similar to modern sea lilies. Their colonies of wispy calyxes were preserved in slabs. In the 1980s, Dad sent slabs of *Uintacrinus* to places as far away as Germany.

When he had visitors in Healy, Dad enjoyed driving them to Shields to see his current fossils. He would race over to Shields, pushing his Olds Delta 88 up to eighty miles per hour on Highway 4 to get a rise out of his passengers. He slowed down to the speed limit to go over the railroad tracks, saying, "Any trains? [*pause*] Only one!" also to get a reaction. Owning the two modest properties grounded him and kept him close to the beds. He worried about both Healy and Shields if he was gone for long. During winter, he was afraid the pipes would "freeze up and bust."

He parked his cars, including a Chevy Nova and the big Olds, on the buffalo grass lawn at Healy. He had no garage, so he drove his best-working vehicle up to the back stoop of the house. In the 1980s, there was a derelict rusty tractor and a third car parked in the yard. He had enough skill as a mechanic to keep the cars running, but technically none of them would have passed state inspection, so he used his artistic skills to paint the current year on his lone Kansas license plate; then he transferred the tag to whatever car he wanted to drive.

Pop was a shade-tree mechanic who could break down cars to their basic parts and fix them, most of the time. He valued the freedom of cars, even though he was usually working on one "infernal combustion engine" or another. He would start the cars he wasn't using, gunning the engines mercilessly, *rudden, rudden, rudden,* then letting them idle for a few minutes. Whenever I hear someone start an old car and rev it, I think of Dad and smile.

On Christmas break in 1982, my sister Chris was finishing her art degree at Fort Hays State, I was attending KU in Lawrence, and Dana and Chuck, both with their master's degrees in art, were living, respectively, in Hays and Liebenthal, a historic German community south of Hays. Dad's four youngest children were near enough to visit him in Healy on holidays and enjoy his famous pot roast with carrots, cabbage, onions, and potatoes. He made a big production of preparing the beef roast in a hefty square pan with a domed lid, blackened on the bottom from years of use. It was actually an old electric skillet. He repurposed it by removing the electric plug section and placing it straight on the gas burner to make his large meals.

Since Dad had taught us the art of making light of everything, Chris and I produced a salute to his religious irreverence that went, "Our Father, which art in Healy / hollow be thy name." The family was sitting around the long dining table at his house, doing what we liked to do most: talking and making jokes. When the roast was ready, Dad called everyone to attention, telling us to get our plates. On the center of the table, he had laid out his well-used (and never washed) potholders. He carried in the roast pan, placed it on the potholders, flamboyantly swept off the lid, and hollered, "Eat, ya hogs!" Everyone laughed—it was one of his running jokes, but we knew how much effort he had put into this holiday feast.

As we filled our plates, he said, "You four are my post-World War II children. You all look at the world so differently from my pre-war children." That proclamation got us loudly debating how we were similar and dissimilar to our four older brothers and sisters. The upshot of the conversation was that the younger crowd's experiences were colored by the war in Vietnam, the hippie lifestyle, and different technology, music, and attitudes. The older set were affected by Dad's service during World War II. They suffered trauma when "Daddy went to war"[4] but they also experienced his post-war prosperity, something the youngest of us did not. And now that all his birds had flown the nest and money was tight, Dad was becoming a bit of an unkempt hippie himself.

The family discussed the state of fossil brokerage. The early paleontologists, including Dad, had all sold specimens to museums but rarely to private individuals. To his chagrin, with the growing popularity of gem and mineral shows and the awareness that there was money in dinosaurs and other fossils, amateur hunters were mining fossil-rich areas. Dealing in fossils was becoming more common. At this point, due to his own tough times, Dad was trying to sell for more than gas and plaster money, but he preferred to stick with museums, he said, "to further the cause of science." Locally, interest was piqued partly because he was getting more publicity, which was causing an uptick in tourist paleontologists. Chuck called them "grab-and-go" fossil hunters. And Dad worried that the interlopers would create tension with the landowners.

Even though he was no longer a member of the Society of Vertebrate Paleontology (SVP), Dad believed in that organization's bylaws on the ethics of fossil collecting, which state:

> The barter, sale, or purchase of scientifically significant vertebrate fossils is not condoned, unless it brings them into, or keeps them within, a public trust. Any other trade or commerce in scientifically significant vertebrate fossils is inconsistent with the foregoing, in that it deprives both the public and professionals of important specimens, which are part of our natural heritage.[5]

After dinner, under a light fixture dimmed by a few dead moths (we called them "millers"), Chris and I started leafing through the family photo albums, chuckling at shots our Mom and our oldest sister, Clare Jane, had taken to document the growth of the children. Almost every spread of the albums had a photograph of some fossil. "Why is there a horn on that fish?" Dana asked, looking over our shoulders at a snapshot of a near-perfect *Ichthyodectes*. *Ichthyodectes* ("fish biter") was a slightly longer, thinner relative of *Gillicus*. It could get up to nine feet long, had round, sharp teeth, and is presumed to have been a swift swimmer similar to a modern barracuda.

The Ichthyodectes that grew a horn, one of the most complete Ichthyodectes ctenodon specimens ever found, is featured in a Sternberg Museum exhibit of Kansas fishes.

"I have no idea why George did that," Pop replied. He said he had asked George Sternberg why he painted a horn on the skull of the best, most complete *Ichthyodectes* Dad had ever found, and George had responded, "I thought he could use a little decoration. It's the holidays, after all."

We all had a good laugh at that, remembering George Sternberg's wry sense of humor. The devilish side of him probably wanted to start a debate among the degreed scientists, an idea that would have amused him. At the time, our father didn't do anything to correct his mentor, playing along with George's flight of fancy. "I sure do miss him," he said, remembering his joie de vivre. "He was quite a George."

"What were your other best fishes, Dad?" Chuck asked. "We know about the large specimens, but among the smaller ones, which were you really proud of?"

"That little *Saurodon* I sent to Shelly Applegate was a dandy," Dad replied. "Clare Jane and Dana both found perfect little *Albula*s,[6] but for some reason the middle-size fish were rarely articulated. We usually just find their scattered fins and vertebrae. When they died, it must have been easy for the scavengers to feed on them and scatter their bones. So, when you get one that's almost perfect, that's really a thrill."

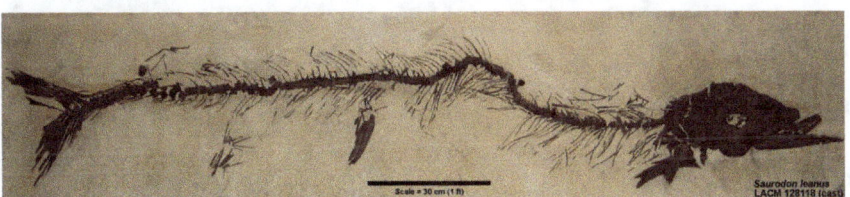

One of the most complete Saurodons ever found. Photo courtesy of the Natural History Museum of Los Angeles County.

The complete *Saurodon* was only one of many high-quality fossils Dad sent to Dr. Shelton P. Applegate at the Los Angeles County Museum in the 1960s. (Shelly, as Dad called him, would become another close paleontologist friend—more about him later.) *Saurodon* were thin predators that could grow up to eight feet long. They had a sword-like lower jaw that they may have used for thrashing or sifting along the sea floor. Complete *Saurodon*s are exceedingly rare, and Dad's specimen that he sent to Applegate is still likely the most complete one collected.

After a time, Dad changed the subject from fossils to literature. He began declaiming lines out of choice books for our benefit. He would recite favorite passages from his favorite novels and poems, quoting such manly authors as Robert

Burns, Zane Grey, Robert W. Service, Jack London, and Rudyard Kipling. His other sources were the King James Bible, Shakespeare's plays, *TRUE* magazine, old movies, and obscure books like W. E. Webb's *Buffalo Land* and Ernest Fletcher's *The Wayward Horseman*. Fletcher's humble book, about an early cattleman in Colorado, was amusing because the author was so self-deprecating. One of Dad's favorite lines from it was, "When we left Woodruff the next morning, the little Mormon kids threw rocks at us."[7]

Dad was right that the pre-war kids had a different upbringing. From my perspective, the older siblings functioned somewhat like parents to me after Mom's death, which occurred the month before my tenth birthday. They helped Dad by letting me stay with them during the seven summer breaks when I was growing up in Leoti. In addition, I learned many details from him about their upbringing—about what it was like to parent these "parents" during the War years and the prosperous post-war years. These experiences fed my father's identity, and he reminisced constantly.

After Mom died, we younger children became his sounding boards. Where he was driven and active with the pre-war group, he was more contemplative and laid-back with us. That may have been a function of aging.

Old Man of the Army
1940s

EVEN THOUGH WE POST-WAR children didn't experience the World War II years, we heard all about them from Pop and our other siblings. Dad's relationship with his brother Jennings was very close, particularly during those years, when Jennings managed the theater while Pop was away in the Army. Dad told us plenty of stories about his earlier exploits with "Jenna Boy" when they were growing up in Wichita County.

As youngsters, Dad and Uncle Jennings went to carnivals in larger cities like Garden City, Dodge City, Wichita, and Denver, Colorado. In the 1920s, there were "freak shows" with such exhibits as the Ossified Man and the Bearded Lady. Carnival barkers added many phrases to Dad's repertoire. Performers who challenged carnival-goers to take part in physical exhibitions (offering things like the chance to step in the ring and spar with them) screamed phrases like, "Anybody here wanna *box*? Anybody here wanna stand on my neck?"

When the two teenagers strolled along, the hucksters aggressively peddled their wares or games, but the loudest shouts came from the food vendors: "Hey! Don't you ever eat, you skinny rascal!? ...Chicken! Chicken! Spring chicken! Chicken without spring! A loaf of bread, a pound of meat, and all the mustard you can eat!"

Dad and Jennings had boxing gloves and sparred regularly, so they were extremely interested in watching the "prize fighters." One time, at a fair in Wichita, Jennings answered the carnival barker's call of "Anybody here wanna *box*?" Dad was extremely proud that his brother "stepped into the ring with Jack Dempsey" in what was undoubtedly a demonstration sparring match. (Dempsey, known as the Manassa Mauler, was from Manassa, Colorado, directly south of Alamosa near the New Mexico border. He held the world heavyweight boxing title from 1919 through 1926.)[1] In another outing in the early thirties, Pop told me, Jennings went for three three-minute rounds with Mickey Walker, the "Toy Bulldog," a world-champion middleweight boxer. Jennings received one hundred dollars for lasting three rounds with Walker.

Boxing was a popular formative sport and an influence in our lives as well, particularly during the years when farm families and townies had only a radio for home entertainment. Even after the advent of television, which we were slow to acquire, we sat together and listened to the bouts. I remember Chuck, Dana, and I listening to the broadcast of Muhammad Ali (then Cassius Clay) defeating Sonny Liston. Dad loved Clay's repartee, seeing in him a fellow character.

Our identities are colored by incidents and memories, and some feel the need to share past experiences to entertain others. Dad was one of those people. Because of the tendency to do some carnival barking himself, the Alpha Dog of the family kept everyone captivated. When it came to listening to such tales as the carnival experiences, we all fell quiet. He was happiest when he held the floor, and we loved the stories.

On September 16, 1940, the United States government enacted the Selective Training and Service Act. This required all male U.S. citizens between the ages of twenty one and forty-five to register for the draft. It was the first peacetime draft in U.S. history. At the age of twenty-nine, Dad registered for the draft but thought he would not be called to serve. After all, he was almost thirty, married, with a growing family.

World War II took place from 1939, the year Hitler invaded Poland, through 1945, the year the U.S. dropped two atomic bombs on Japan. The Kansas Geological Society summarized the effect of the war on Kansas:

> By 1945 and the end of the war, nearly every community in the state, no matter how small or remote, had sent someone to the war. ... Several Kansas towns became homes to newly built airfields—huge landing strips that were training bases for pilots before they were shipped overseas. The larger cities, such as Wichita and Kansas City, experienced manufacturing booms, especially related to aircraft. Commodity prices increased and improved the income of those who stayed home.[2]

Rising commodity prices benefited farmers in all Kansas counties. Since farmers were considered essential to America's war effort, they would be, by and large, exempt from the draft.

The motion picture industry cooperated with the U.S. Treasury to push bond sales in theaters nationally. As manager of the Plaza Theater, Dad helped with the war effort as chairman of the bond drive to raise money for war efforts. Citizens of the World War II generation remember liberty bonds and war bonds. Purchasing them became a patriotic gesture and introduced many Americans to financial securities for the first time.

In his later years, Dad enjoyed the dramatic effect of calling his first four children his pre-war family and his last four the post-war family. Orv and Clare Jane were born in 1936 and 1938, respectively, and my parents had two more children in the early 1940s before his Army service. While the first two were named for O. W. and Clare, the third child, Violetta Jo (Letty), born in June 1941, was named after Grandmother Viletta Markley Bonner. Mom inserted an "o" in the first name on the birth certificate, so the two names were spelled differently. Perhaps Mom was influenced by the Louisa May Alcott novel, *Little Women,* in choosing Jo for Letty's middle name. Two and a half years later, in December 1943, Ralph Stephen Bonner (Steve) was born. He was named after Berg relatives—Uncle Ralph and our maternal grandfather, Ralph Harvey Berg. His middle name, which became what everyone called him, came from one of Dad's beloved Nebraska uncles, Steve Bonner.

Dad unconsciously repeated the bonding with his oldest son the way his father, O. W., had done with Eldredge. Orvie was Dad's boon companion, and Mom and Dad encouraged Clare Jane and Orv to help with the babies, instilling responsibility in them but also giving them a premature sense of adulthood. When five-year-old Orvie found out his parents were expecting their third child, he asked, "How much is this baby going to cost?" Dad answered, "Fifty dollars." Astounded at that sum, Orvie advised, "Don't you think you ought to quit having babies and finish the theater?"

The Bonner matriarch, Grandmother Viletta, died in 1939, the same year the Bonners purchased the former Leoti Auditorium. In addition to renovating it into the new Plaza Theater, they added offices and apartments on each side of the long central auditorium section. Mom and Dad fixed up a flat on the south side of the theater building for O. W., now a widower, to live in until his death in 1951.

Dad, Mom, Orvie, Clare Jane, Letty, and new baby Steve occupied the larger of the two apartments. With many family members in close proximity, Mom fed everyone, including O. W. and whatever other relatives visited. Dad called it "feeding the multitudes," implying that she was a miraculous provider.

Steve was a plaything for Clare Jane and Letty, who were modeling Mom's mothering skills. One time, Mom wheeled Baby Steve to the grocery store in a rattan baby buggy. The two little girls came along, each of them holding a large doll. When Mom went inside to get her groceries, she instructed the girls to keep an eye on their little brother, who had fallen asleep on the short ride.

"My, what lovely dollies you have," said a dowager who was walking into the store. Both girls looked at their dolls, which were blonde and brunette, like them. They nodded at the well-meaning woman. "You have a sweet little dolly in the carriage, too," she said, smiling. To which the three-year-old Letty answered, all-knowingly, "That's not a dolly, that's Ralph Seeben!" The lady snickered and walked into the store. Clare Jane repeated the interchange to Mom, who praised Letty for her wisdom, tongue-in-cheek.

During warm weather, the whole family, even the baby, went to the fossil beds. Our parents were so busy with their growing family and working in the theater that these visits were infrequent, a weekend picnic every other month. But they believed it was important to take their children to the "fossil bed playground," to give them an appreciation of science and nature.

For years, Dad didn't pay much attention to invertebrate fossils. Broken seashells were one of the first things we children picked up, but we soon learned that Dad would not be spending time collecting these organisms. Small pieces of *Ostrea* were scattered in many of the fossil beds. These mollusks were commonly called *Ostrea congesta*, which means "heaped oysters," but the current recognized name is *Pseudoperna congesta*. Pieces of these oysters were more plentiful than shark's teeth, fin material, and random vertebrae. Early Kansas paleontologist Benjamin F. Mudge noted that the most common mollusks in the Niobrara were *Ostrea* and *Inoceramus*.[3]

B. F. Mudge (1817–1879) established the Geological Survey of Kansas in Manhattan. Considered by many as the founder of Kansas paleontology and geology, Mudge was a lawyer, educator, and geologist. He was the first State Geologist of Kansas and co-founded the Kansas Academy of Science. Mudge sent most of his fossil discoveries back East to Cope, Marsh, Agassiz, and other scientists. He was the first to discover the toothed bird *Ichthyornis* in western Kansas and sent the type specimen to O. C. Marsh. He also collected large dinosaurs, such as *Allosaurus* and *Diplodicus*, in Wyoming.

Uncle Jennings scrambled up the ladder to the projection booth where Dad was oiling the projector. "Did you hear?" he yelled over the whirring and whacking noise of the machine. "The Japanese bombed Pearl Harbor! I just heard it on the radio."

Dad turned off the projector and gawked at his brother. "Where the hell is Pearl Harbor?" he asked. Jennings explained it was a U.S. naval base in Hawaii. The news, sinking in, was chilling. Aware that the Axis powers were Japan, Italy, and Germany, Dad exclaimed, "This is an attack on our country! What's going to happen now?" The two brothers hurried to the small apartment on the side of the building to tell Mom the news. The three of them sat together, drinking coffee, realizing that the U.S. could not remain neutral now.

At the time he registered for the Selective Service, Dad felt relatively sure that he would not be drafted. But now, after the catastrophe of Pearl Harbor, he wasn't so confident that he would be spared. As the news came in, the family stayed close to the radio to keep track of what was going on. Dad ran newsreels before each feature film, and those short movies were setting the stage for American immersion into what was becoming another world war. It became clear that many young men would volunteer as well as be drafted to serve their country and fight against the Axis powers in Europe and Asia.

Now that he was an adult and a father, Pop was no longer a farmer. Because he worked primarily in what the government deemed a nonessential industry, the movie theater, Dad was swept into the larger net of the draft after Pearl Harbor. He received his "Greetings" letter from Uncle Sam the year he turned thirty-three.

The *Leoti Standard*'s front-page item, "Three Selectives Leave Leoti" on February 3, 1944, described a procedure in which the men were sent to pre-induction physical examinations. Once examined, the men would either be sent to the Army, the Navy, or rejected. "The second group of men to be sent to Fort Leavenworth from Wichita County, for this pre-induction examination under the new regulation, left Leoti Sunday night. The group included: Marion Bonner, Lee Daniels, and Tod Heath. [The] three men passed their pre-induction physical examinations and will leace [*sic*] for the replacement center after 21 days."

Dad, Mom, and the kids went to the fossil beds one last time before Dad left for basic training. On this trip, he found a partial *Enchodus* fish, which he didn't collect because it had only some fin material and a few scattered vertebrae. Most of the species of *Enchodus* were sardine size and smaller. One species of *Enchodus* ("spear tooth") in Kansas reached four to five feet (*Enchodus petrosus*), and that is the version Dad knew about. At the front of its jaws were fierce fangs to attack small prey, but it also served as prey for larger fish and reptiles. That day Dad also found a couple of shark's teeth and a nice *Gillicus* vertebra to take with him as a reminder of home.

The Bonners with their four pre-war children outside the theater building shortly before Dad's induction into the Army. Mom holds Steve. Front, left to right: Letty, Clare Jane, and Orville.

Our father left home in early 1944, and on his birthday in May of that year, he turned thirty-four. Although having four children had no effect on his draft status, his age may have kept him stateside. He served in the Army from February 1944, two months after Steve's birth, to December 1945. Japan's surrender to the U.S. occurred in August 1945, but it was not until Christmas of that year that Dad returned to Leoti.

He took his basic training at Fort Bliss in El Paso, Texas, where he ran movies for the troops in one building and for the officers in another. He earned a projectionist's permit, prosaically titled, "United States Army Projection Operator's Permit" from the Signal Corps Training Film Library on August 31, 1944. Because of his age, the other soldiers in his barracks, hailing from far-flung states like New Jersey, Mississippi, and Nevada, called him "Old Man." They entreated, "Hey, Old Man, want to play some poker?" Or "We're going to Cuidad Juarez tonight, Old Man. Come with us!"

During the war, many fathers had to leave their families behind. The biggest concern, of course, was whether the fathers would see combat. Fortunately for Dad, by the time he went through basic training and received stateside assignments, his age and status as a parent of four worked in his favor. But nothing was certain. When fathers left, whether in stateside posts or overseas, separation and absences saddened and worried families. "Psychologists have shown that of the children who suffer separation and absence, none are more affected than the little children. This was certainly true for America's homefront boys and girls born during the war."[4]

Before Dad's induction, Orv, the "little man" (Pop's term) went around Leoti with a wagon, picking up bottles and anything else he could sell. In 1942, at the age of six, he broke his leg when he fell from swinging off an awning. Doc Ott had his office across the street. Mom, in the theater apartment, quickly spirited Orvie over to the doctor. Doc Ott put a cast on Orv's leg from ankle to mid-thigh and said to Mom, "That'll be five dollars, Mrs. Bonner."

When Pop left home in 1944, Mom's brother, Ralph, was already in the Army. He was blind in one eye, but the Army accepted him into the Signal Corps, and he spent many years stationed in China. Now Mom and Grandmother Berg needed to cope with both Ralph and Dad being in the Army. The stoicism of women like my Mom and Grandmother was a necessary attribute during the war years, when wives and mothers knew they must hold down business at home. They accepted the fact that Dad had been drafted while still preparing the young family for the separation. In their minds, he wasn't going overseas; therefore, the fear of combat was allayed.

Even so, children nationwide encountered the idea of death in war in diverse ways, often from the movies. In the 1944 film *The Fighting Sullivans,* all five of the Sullivan brothers were killed in the Navy, with the movie playing up the themes of heroism and self-sacrifice.[5] Most likely, the Bonner family and theater audiences were informed by the newsreels that preceded each motion picture. They also absorbed pro-American propaganda from such shows as Irving Berlin's 1943 hit *This Is the Army,* with its well-known song, "This Is the Army, Mr. Jones," where raw recruits celebrate their involvement in the "greatest Army in the world."

The birthday cards Mom saved from Dad during their separation spoke volumes. He missed two of her March birthdays while serving his two-year stint in the Army. In 1944 he was in basic training at Fort Bliss in El Paso. The March 21 card, purchased from "slim pickin's" at the post exchange, has a pink floral landscape scene. On the back, he wrote, "Happy Birthday Sweetheart! I couldn't find anything decent over to the P.X. to send you so will send you this card, it's dirty too. Ha. But you know my feelings are all for you this day and every day as long as I live. Of course, you won't get this until after your birthday but better late than never. Ha. I'll write you tomorrow. Your Sweetheart, Marion."

A year later, there were more goods available, so the card accompanied a gift. The 1945 card, sent from "Pvt. Marion C. Bonner, Co. A, 55th I.T.B., Camp Howze, Texas" showed a scene of bluebirds flying around a nest. He wrote, "Happy Birthday, Sweetheart. Hope you like the earrings, Honey. Couldn't find any to match the stone but these will go with anything. Here's hoping I'll be with you for your next birthday, Honey. Your loving husband, Marion."

Fort Bliss was a huge, well-established post, with hundreds of brick buildings and multiple motion picture theaters. Dad operated all of them. After basic, he served for a few months in Camp Howze, Texas, an infantry replacement training center near the north Texas city of Gainesville. According to Orville, "When he got to Camp Howze, his commanding officers said, 'I see you ran theaters at Fort Bliss,' and he answered, 'I'm kind of sick of running theaters,' so they put him on KP."

While at Camp Howze, he learned he would be transferred to Arkansas. When Mom heard Dad was moving to Camp Dermott, in southeast Arkansas, a remote facility that housed German prisoners of war, she was apprehensive. Then she discovered that they would receive a family military stipend and live together in a modest duplex that was a government building made over into a residence. The family went to Arkansas in 1945 after school was out, effectively becoming "camp followers." Had the war continued, they would have likely followed him to the next assignment. Like many children of the era, the Bonner kids saw life as they knew put on hold until the war was over.

The trip from Leoti to Jerome, Arkansas, by way of Wichita and Tulsa, Oklahoma, was more than eight hundred miles in the black Studebaker. It took them three days, and they stayed one night in Wichita with Aunt Veda. As soon as they got on the road from Veda's house, the car broke down. They limped into Eureka, Kansas, sixty miles east of Wichita. There Dad called the Arkansas camp and explained the problem so that he wouldn't be counted as AWOL. The family stayed in a hotel in Eureka while the car was patched together. The groaner of a joke in later years was that Christine, their fifth child, was conceived in Eureka!

The rustic duplex in Jerome, shared with another family, sat in a woody area much different from the wide-open plains of western Kansas. The setting was fun for the kids because they made friends with a gaggle of neighbor children—also displaced from their homes—and were able to run and play in the wild areas around the house. They had a fun time making up games under the pier-and-beam structure.

Clare Jane was seven years old and attended the second grade in Arkansas. Orville was nine, entering the fourth grade. Letty was four, and Steve was a toddler. Some of Letty's first memories are of the house in Jerome. She says that with Dad working at the prisoner of war camp, Mom held the family together in the two-story duplex. "Mom was simply wonderful. She always had us doing fun things, and I remember watching her paint a tree-filled landscape that was outside the door." Orv still has the painting. The small home had a pot-bellied stove for warmth and a hot plate for cooking, which Mom managed very well.

When my brothers and sisters were asleep, Dad and Mom, who was pregnant with their fifth child, spent the nights planning what they would do back in Leoti. On a private's pay of twenty-five dollars per month, he had not been able to save much, but he had put aside a stash of poker winnings. It wasn't hard for the Old Man to earn a few bucks from the other privates, but he really scored when he beat the sergeants, who earned a higher salary and had more money to lose. Mom's family stipend was one hundred twenty dollars a month, and since she was used to making do on much less, she saved some cash too.

U.S. Army portrait of Dad, 1944.

They discussed the possibilities of buying land, buying a lot in Leoti, and building a home. The theater apartment was too tiny. While they were in Arkansas, they started talking about their dream home. Both were intrigued by the Prairie-style architecture made famous by Frank Lloyd Wright. They bought a book of his home designs and flipped through it, envisioning what would be possible in the small town of Leoti. The seed was planted for their post-war American Dream.

Coming back from the Arkansas sojourn in the summer of 1945, Orv and Clare Jane were excited to know they would be starting the next school year back in Leoti. For the return trip, the family again planned to stay over with Aunt Veda in Wichita, and they all piled into the Studebaker. The car, loaded with kids and household items, sputtered and creaked all the way home. When they pulled up beside the Plaza Theater in Leoti, the Studebaker's driveshaft dropped out. I cannot say what that means, exactly, but it must have been dramatic because the phrase was repeated over the years.

Because Dad wouldn't muster out of the Army until the following December, Mom and the kids stayed with Grandmother Berg in her small house on Fifth and Logan streets in Leoti. She provided hours of childcare and tutelage for the children in the 1940s. When she wasn't working in the library, my oldest brothers and sisters would spend quality time with her, learning bridge, hearts, rummy, and Chinese checkers. Importantly, she provided advice, strong moral support, and comfort to our mother.

On an unseasonably warm day in January 1946, a month after his return to western Kansas, Dad heard the call of the chalk. He now drove a white Chevy sedan (the Ghost) that he bought to replace the Studebaker. He urged the car carefully over the pasture, paying close attention to the rutted road used by the ranchers. He wasn't as careful once they got near the outcroppings, which shone in the sun like the exposed layers of a tumbled-down yellow birthday cake. He pulled the car close to the edge, approximately four feet from the canyon's drop-off, grinning devilishly at his pregnant wife.

"Where angels fear to tread," she exclaimed loudly. He beamed and she smirked, glancing over the head of Steve, the child sitting between them in the front seat. It was their little inside joke. The kids all heard a bold compliment, but Dad and Mom were both literate enough to know the first part of the Alexander Pope proverb was "Fools rush in." It was her way of expressing concern that he'd gotten close enough to the canyon drop-off.

"Where angels fear to tread!" Dad echoed expansively. He turned off the ignition, and the four kids tumbled sleepily from the car. Dad and Mom were running the Plaza Theater every night, so their only way to get in any fossil-hunting time was to get to the chalk early, hunt past lunchtime, then return to Leoti by late afternoon to prepare for the movie screening.

The kids gathered at the back of the car, where Dad handed each of them their fossil tools. The tools they received were related to their birth order and to how serious they were in the pursuit of fossils. Orvie, who had just turned ten years old, received a small awl for close digging, an old paint brush, and a Marsh pick. Clare Jane and Letty both got awls and brushes and even three-year-old Steve received a small brush. This was a ceremonial ritual Dad reserved for his wife and children. No one grabbed a tool. They all waited as tools were distributed. As the children got older and more responsible, if they planned to go far, they took a small canteen with them. The older kids knew how to examine and gently dig and brush around a fossil to determine if it was worth excavating.

That day, Letty found what she said was her best fossil, worthy of collection. It was a perfect fin of a *Protosphyraena*. *Protosphyraena* ("early barracuda") were similar to modern swordfish, with a pointed snout and long pectoral fins. Vertebrae of this fish did not fossilize, and there have been very few tails found. The fins are prevalent, but material sometimes attributed to *Protosphyraena* has since been determined to be scattered *Bonnerichthys* remains.

In addition to finding fish fins like the beautiful *Protosphyraena*, Letty remembers always looking for shark's teeth and conical mosasaur teeth. Dad instructed everyone to not just pick up teeth and pocket them but to look around carefully to see if there were other parts of the skeleton nearby. All of us learned to mark the spot with a pile of rocks and a big X carved into the chalk wall.

Fossil bed education was recreational. War education was much more serious. The way they learned about the war, like other U.S. children, was through film. In addition to escapist films like musicals and comedies, "at some theaters the children viewed classic films funded by the government, notably the documentaries in Frank Capra's *Why We Fight* series as well as [movies such as] John Ford's *The Battle of Midway* (1942)." Children also saw "government-sponsored public service films and cartoons; the best-known of these cartoons was Walt Disney's *Der Fuhrer's Face*, starring Donald Duck, which gave the nation a hit song based on the unlikely topic of flatulence: 'We heil [Bronx cheer], heil [Bronx cheer], right in der Fuhrer's face.' "[6] Even the younger brothers and sisters repeated this song, long after the war. I remember singing along with the funny chorus.

Leoti established a war memorial for Wichita County veterans in 1951. It was a renovated World War II cannon, placed on a sturdy concrete plinth in the courthouse square, directly across the street from the Plaza Theater. The piece of artillery was a 155-milimeter howitzer used by American forces during the war. Installed below the cannon on the sides of the concrete base were bronze name plates listing Wichita Countians who were killed and who served in World War I and World War II. Nine men gave their lives during World War II, and 268 served, Marion C. Bonner listed among them. When she was a student at Leoti Junior High School, a spunky Letty Bonner surreptitiously placed a dot of chewing gum beside our Dad's name. It was there when we younger kids were growing up in Leoti but got smaller and smaller over time and is now gone.

The wartime experience of Kansas was similar to that of rest of the states. Summarizing the war's effects, historian Kenneth S. Davis wrote, "Some 215,000 Kansans served in the armed forces. Of these, 3,879 were officially listed as killed in action or dead of wounds."[7] But Kansas was affected positively in two major ways: an uptick in income after the Depression (because of the demand for farm products, whose prices skyrocketed) and a psychological surge in pride due to the crucial role of fellow Kansan Dwight D. Eisenhower.[8] The five-star general of the Army was Supreme Commander of the Allied Expeditionary Force in Europe and was propelled to the presidency due to his wartime popularity. After the war, even staunch Kansas Democrats like my parents and Uncle Jennings were heard to say, "I like Ike!"

After World War II, Jennings was ready to hand the management of the theater back to Pop. And Dad was ready to become "bull of the woods" again. The two brothers shared phrases and lingo that was sardonic, self-mocking at times. Dad and Jennings sounded alike, but their personalities were strikingly different. Dad was bombastic, but Jennings was a kind-hearted listener who put in his witty two cents when he could get a word in edgewise.

Jennings served as Register of Deeds and County Treasurer in Leoti in 1934 and 1940 but stepped down when he was denied days off to help with the wheat harvest. Like Dad, Jennings was fascinated with the past, but his focus was on human history, not prehistory. Even though he was in the group of boys who went fossil hunting with science teacher Arthur Wedel in 1925, Jennings didn't catch the paleo bug the way Dad did.

Uncle Jennings eventually moved from Wichita County to Colorado. There he was an assiduous amateur archaeologist and historian of the Old West. A chaser of dreams, in the 1960s, Jennings owned the Wonder Tower in Genoa, Colorado, a tourist stop near Denver. In 1932, the tower was recognized by "Ripley's Believe It or Not," which claimed that visitors could see six states from its top. My older brothers and sisters remember visiting Uncle Jennings in this marvelous place, going up its narrow, circular stairs to view the Great Plains from a height of sixty feet above a town that is called the highest point between Denver and New York.[9]

In the 1970s, Jennings opened a curio shop and antique store he named Black Kettle Lodge in Chivington, Colorado. Colonel John Chivington was the man who in 1864 led a force of volunteer Colorado Cavalry to annihilate a village of Cheyenne families camped along Sand Creek. Eminent historian Robert M. Utley describes Chivington, the Fighting Parson, as a man who earned fame in battle with Confederates but is "rightly stigmatized for his treacherous attack on Black Kettle's Cheyennes at Sand Creek."[10] Black Kettle was a peaceful chief who represented the last hope of the Cheyenne.

Uncle Jennings's Black Kettle Lodge, a last hope of a business, was a long white building containing an antique store and curio shop on the east end and Jennings' apartment on the west end. The one-story structure sprawled beside a bend of Highway 96 in eastern Colorado, sixty-five miles due west of Leoti.

Trucks roared around the bend, and cars rarely stopped. As a teenager, my brother Chuck painted the likeness of Black Kettle on the side of the storefront.

Jennings developed a heart problem and closed his final venture when I was in college. He passed away peacefully in Calhan, Colorado, in 1979. Dad felt the loss deeply, and as a family, we agreed that we had never known a sweeter, more gentle man than "Jenna Boy."

Theater Operations
1920s–50s

WHEN O. W. BONNER opened his movie theater in the Santa Fe depot in 1926, popular male leads of the silent movies were Charlie Chaplin, Buster Keaton, and Lon Chaney, the Man of a Thousand Faces. Female leads Lillian Gish, Clara Bow, and Greta Garbo mesmerized audiences. Silent films relied on subtitles to explain the onscreen action. Wichita County residents remember that when the movie theater operated in the depot, the Bonners would leave a water bucket and dipper sitting by the door for all to get a drink.

At the start of their picture show venture, Pappy and Grandma Bonner were fifty-two and fifty years old. Most of their sons and daughters were living in Leoti when the theater replaced farming as the primary income source for the family. O. W. purchased the twelve-foot-wide motion picture screen, projector, and other trappings from Charles Swann.

Wichita Countians enjoyed the entertainment, and the business soon needed a bigger space, so O. W. rented a building across the street from the old Leoti post office. This pattern followed that of many small Kansas towns, where movies were screened in existing buildings on Main Street and "an open interior room with movable seats and a screen attached to one wall formed the auditorium."[1] These venues had moved past the nickelodeon days, and the Bonners now charged ten cents per person.

In the pre-Hollywood era, the Bonners rented movies from distribution companies based in New York. Then Hollywood developed the talkies, and they ordered movies from such early studios as Paramount and Republic. The talkies brought new challenges. In order to have a soundtrack match the action onscreen, they used a method called "sound on disk." Pop told us it was a frustrating experience to try to "put it in sync" if the sound stopped matching the lips of the actors. Warner Brothers was one of the early companies to provide sound on disk. As technology advanced, recorded dialogue made its way onto the film itself, and sound on film became commonly used by the 1940s.

One time Pappy chose a British indie film in black and white and with sound. The production quality of the reels was so bad that the entire movie looked very dark, as if shot in a cave. It was impossible to adjust the projector to see what was going on in the action. Dad remembered that he could see only the tips of the characters' cigarettes when they took a puff. He didn't tell me the name of this movie, but one of its key lines of dialogue was, "I'm damned sorry, Old Man." That line was repeated so frequently that it prompted O. W. to mutter, "And I'm damned sorry I booked it."

During the days of the silents and early sound, O. W. often read the names of the actors when they appeared onscreen, with unconscious malapropism. He pronounced the actor Maurice Chevalier as "Morris Cavalier" and Patsy Ruth Miller as "Pastie" Ruth Miller. In later years, Pop repeated these "Pappy-isms," but it is unlikely he made fun of O. W. to his face. Disrespect for one's elders was simply not allowed. Sometimes Pappy's coinages were clever, though. In the old depot, if a popular film had standing room only, the Bonners put up a sign at the ticket booth that read: "SRO." Pappy informed a confused farmer that this meant "seats run out."

Experiencing many SRO showings in the rental space during the Depression years, the family realized that the theater needed to expand one more time. So Pappy purchased the former Leoti Auditorium in 1939. With the help of local carpenters and electricians, the Bonners remodeled the long, tan stucco building on Main Street, one block south of the intersection of Highways 96 and 25. They bought a marquee that bore the name "Plaza Theater."

Of the seven children of O. W. and Viletta, only the two youngest, Jennings and Dad, as well as their niece, Janice, graduated from Leoti High School. The older sisters and Eldredge had gone to high school in Imperial, Nebraska. And just as the previous generation of Bonners scattered from Imperial, so the next generation dispersed from Leoti. Dad was the only one of his family to spend most of his adult life in Leoti.

By 1940, Dad and Pappy discussed ownership of the theater (Dad had been running it for several years by that point). O. W., now a widower, decided that after his death, his youngest son could buy out his siblings. They settled on a figure of twenty-five thousand dollars for Dad to become sole owner. "We can have you pay my heirs, each of your brothers and sisters—and Janice—an equal share of five thousand dollars," O. W. told him. Dad did the math in his head: Veda, Helen, Eldredge, Jennings, and Janice made five people for five thousand each. "Wait a minute," he said to Pappy, somewhat dismayed. "Don't I count as one of the heirs?" O. W. looked at him in shock, stroked his chin, and replied, "Oh, I guess you do. So split the amount to pay each of them by six, then." Dad agreed to the deal. The inclusion of his niece Janice as one of the siblings seemed odd, but then he reasoned that his parents had essentially raised her, and Pappy wanted to help provide for her future. (To me, this story was an example of my Pop being taken for granted or overlooked by Pappy.)

When the Bonners were ready to open the new Plaza Theater in 1941, they held a grand opening, which they advertised in the *Leoti Standard.* The premiere movie was *One Foot in Heaven,* a biographical film about a Methodist minister, starring Frederic March. Since the seating capacity was nearly three hundred, they didn't have to worry about seats running out in the new location. A story in the *Standard,* titled "Tonight is 'First Night' at New Plaza Theater," explained that "O. W. Bonner and son, Marion, and the entire crew of helpers will see months of hard work come to the finish line today with the opening of the most modern theater in this entire district."

Ownership of the Plaza Theater meant Dad and Mom were completely invested in the venture and the children would start working there too. Two years after opening night, my brother Orv started sweeping up after the shows. He became a projectionist in high school and remembers the theater days fondly: "It was a good business for a time. That was quite a life—the theater at night, and the days for farming and fossil hunting."

As it turned out, 1941 was a monumental year. Letty was born in June, the Plaza Theater had its grand opening in November, and Pearl Harbor was bombed in December.

Growing up, all my brothers did another kind of hunting with Dad—for live game rather than dead fossils. The girls in the family were not given this option, and it didn't occur to us to ask. The boys were not trophy hunters; we ate what they killed, usually ducks and pheasants. Orv says Dad bought him a shotgun when he was ten and that Dad showed him the proper ways for handling and cleaning guns that he learned in the Army. The duo went duck hunting in 1946, a wet year that created ponds or lagoons on fields and pastures throughout western Kansas. All the extra water drew the migration of pintail, mallard, teal, and bufflehead ducks. After the ponds dried up, Dad and the boys went to Eads, Colorado, to hunt at a large reservoir that attracted many species of ducks and Canada geese.

When Steve reached junior high age, the older brothers focused on the area's reservoirs for duck hunting. Later, when the two youngest brothers, Chuck and

Dana, were old enough to hunt, Dad dropped them off at the pastures near the fossil beds that held natural ponds. The Bonner photo albums are filled with pictures of hunters with their game, prior to dressing it out so Mom, and later Dad, could prepare it for dinner.

Hunting larger game was a less frequent activity, but Dad did hunt deer in Colorado with his brother Eldredge in the late 1940s and early 1950s. On one deer hunt in the high country near Pagoda Peak, Dad got lost while following his quarry. He was stranded alone during a heavy Colorado snowfall that started in the late afternoon. He didn't panic but made himself a makeshift shelter by breaking some boughs from a pine tree. Knowing that more snow would fall overnight, Dad decided to hunker down under the boughs rather than continue walking and potentially freeze to death.

The next day, Eldredge sent people out looking for him and hired a tracker to help. Dad came out of his shelter and decided to walk downhill, get to a stream, and follow it. It was easier to see what was going on in the daytime after the snowstorm had passed over. At mid-morning, the tracker said to Eldredge, "Your man is in camp," because he spotted Dad, walking along the stream, making his way back to their rented cabin.

To keep his spirits up while walking, Dad recited poetry to himself, proclaiming verses loudly in the snowy silence, hoping his fellow hunters would hear him. Robert W. Service poems of the Yukon like "The Cremation of Sam McGee" played through his mind and helped him keep the Colorado blizzard in perspective.

An item from the February 14, 1946, *Leoti Standard* read, "Marion Bonner is making great headway with his residence in west Leoti." (Even though the house was only two blocks west of Highway 25, it was considered "west Leoti" then.) At that time, Dad was busy working on his and Mom's first family home on two lots they purchased right after the war. He started on the house when Orv was ten, Clare Jane eight, Letty five, and Steve three. He finished it in 1947, after their fifth child, Chris, was born.

The modest rectangular house at the intersection of M and Second Streets would later be called the west wing, or the Other End. It had high windows on the back end and small multiple-pane windows on the front. On the middle of the south-facing end, Pop installed glass-tile bricks. He built this house on the west end of the lot with the intention of constructing its mirror image on the east end of the lot at a future date. Dad installed screened louvers under the front windows. That was an innovative idea to allow the house to be ventilated during warm months.

Soon after the house was built, a runaway dog adopted my brothers and sisters. He was a pit bull terrier mix they named Rex, and he looked a lot like the RCA Victor mascot. His self-appointed duty was to shadow my brothers and sisters everywhere they went. There are numerous pictures of this loyal dog in our family albums, being hugged or standing next to the youngsters. Rex was an alpha male, always fighting with other roaming town dogs. He was also obsessed with chasing the family car. More than once, Mom accidentally hit him while he was charging the rolling tires. It made a sickening *fluff fluff fluff* noise under the car's chassis. When this happened, Rexie crawled to the side of the house, lying low while his wounds healed. Despite the repeated trauma, he lived to the age of sixteen, passing away in 1962.

The pre-war children stand outside the newly built home on M Street in 1946 with their rescued dog, Rex. Left to right: Steve, Letty, Clare Jane, and Orville.

While the first four children were named for our maternal and paternal grandparents, Mom and Dad's fifth child, Marion Christine Bonner, born in February 1946, received her first and middle names from our parents (Christine was Mom's middle name). The third daughter became, like Steve, another baby for the older girls to look after and play with. Baby Christine, or "Bisteen," was soon simply called Chris.

At the time of her birth, Mom and Dad were enormously busy. Dad was juggling wheat farming (which he returned to, on his own land, after the war), running the theater, and fossil hunting. Mom played a significant role in the theater, helping with bookings and finances while handling the household. Clare Jane remembers that in that trusting era, Mom sent her home from the show, from second grade on, with a bag containing the box-office take for that night. Our house was three blocks from the theater. Clare Jane always ran quickly home and placed the money in Mom's dresser drawer. Having endured the Depression and war years, Mom always stashed away money for a rainy day. One time Dad found a shoebox full of silver dollars stored in the back of the closet. Sometimes when Mom showed her extremely frugal side, Dad ribbed her and called her "Adolphus" after her Grandfather Doyle, but he appreciated how resourceful she was.

In addition to working in the theater, managing the growing family kept Mom on her toes. She took all of us to Sunday School at the Presbyterian Church in Leoti. She cooked huge Sunday dinners, with Pappy sitting at the head of the table. For occasional getaways, my family would take the 270-mile-long trip to Wichita to visit our aunts Veda and Helen. Aunt Helen, now a widow, had purchased a boarding house in Wichita on Park Place that had plenty of room for company.

When in the big city of Wichita, my brothers and sisters enjoyed walking around downtown, window-shopping, and visiting parks and playgrounds. Joyland, an amusement park, opened in Wichita in 1949, and we visited there often—even into the years when I was a preschooler. Everyone loved the Ferris wheel, wooden roller coaster, and painted carousel. Two longstanding attractions at Joyland were its miniature train and Louie the animatronic clown, who played a huge Wurlitzer organ that echoed throughout the park. Joyland's carousel is still running in a botanical garden in Wichita.

Closer to home, every month or so the family spent Saturdays in Garden City, the nearest town of any size. It was only sixty miles from Leoti, sixty-five if Dad took "the river route" through Lakin instead of Scott City. Every trip to "Garden" involved shopping, and each child could purchase a cheap toy from Woolworth's, such as a paddle ball, jacks, or a magic eraser tablet. Pop continued this tradition into the 1960s for the younger half of the family as well. Dad always bought a bag of the warm salted cashews that sat under lights inside the glass counter for his small humans, but he also picked up unsalted peanuts in the shell for Penny, the elephant at the Garden City Zoo. I was amazed at how Penny always remembered Dad, plodding over toward him to get her treats. If we went to Garden the Lakin way, we drove past Holcomb on Highway 50 and gawked at the Clutter house (made famous by the murders Truman Capote documented in his true crime book, *In Cold Blood,* in 1966). That was chilling.

Mom and Dad posed in a photo booth during one of the family's visits to Wichita, circa 1948.

When my older brothers and sisters got to be nine or ten, after each movie showing, they swept the rows of seats to the center aisle, then swept the aisle down to the bottom. This was the pattern Orv, Clare Jane, Letty, Steve, and very briefly, Chris and Chuck followed. When they reached high school age, my older brothers ran the projectors and my sisters popped popcorn. Mom usually handled the ticket sales, but Clare Jane and Letty did a little of that too when needed. Heavily buttered popcorn is a favorite family treat to this day. In later years, Dad popped "good corn!" in Crisco, in a large aluminum pot. Popcorn served with orangeade from a carton, which he called Nehi, always brought back memories of the show.

Letty remembers spending almost equal time between school, home, and theater. She, like Clare Jane, took her responsibilities seriously, but she also enjoyed cutting up with the public while waiting on them at the concession stand. A natural entertainer, she was in many ways a female version of Dad. From Dad and from the films themselves, Letty learned the joy of comedy and performance.

The sixth child, Chuck, was born Charles Randolph Bonner in May of 1950. He received one name from our parents (Dad's middle name) but "Randolph" came from the actor Randolph Scott. A precocious youngster, Chuckie was passed around between his parents and sisters and also learned to be a little performer. Some of his expressions made outside the Bonner home gave people a start. In Sunday School, the teacher asked the children to draw pictures of what they were thankful for. Chuck's tableau involved prehistoric life, and at the top was the heading, "Thank you, God, for Dinosaurs." One other time, the teacher asked her small students to go around the circle and say what they appreciated. Chuck's little friend said in a nasal voice, "A new baby sister." When it was Chuck's turn, despite the fact that he had a new baby brother, he hollered, "Cowboy shows!"

With a life encompassing the theater, farming, high school activities, sporting events, church events, and clubs, getting out to the fossil beds was rare but valued. A trip to Logan County created a short breather for our busy family. In 1957, Clare Jane found a beautiful, complete *Apsopelix* (at the time, Dad and George Sternberg called the fish *Leptichthys*). This fish is at the Sternberg Museum. *Apsopelix* was a small fish found in the Smoky Hill Chalk, measuring about twenty inches long. It was a favorite food of larger predators and probably fed on plankton. This fish is often preserved whole and has dense scales. *Apsopelix* has a confusing taxonomic history, with several earlier genuses, such as *Leptichthys* and *Syllaemus* currently placed in this genus.

Another complete fish that Dad delivered to George Sternberg was a *Pentanogmius* that he collected in 1956. This genus, similar to a sailfish, was substantially larger than the trout-sized *Apsopelix. Pentanogmius* ("bonefish") had a deep, three- to five-foot-long body similar to that of a modern angel fish. Its small, comb-like teeth and multiple palatine plates excelled at crushing invertebrates such as clams and mollusks. The Bonner specimen at the Sternberg Museum is complete except for the tail. Dad always called these sailfish *Anogmius*, based on Cope's name for them. Decades later, the name changed to *Bananogmius*, then to *Pentanogmius* in 2000. In *Oceans of Kansas,* Michael Everhart explains, "While *Bananogmius* was the most common genus of plethodid in the Smoky Hill Chalk, its most common species has now been placed in another genus … specifically excluded from the original genus and placed into a new genus, *Pentanogmius.*"[2]

Many years later, in 2008, my brother Chuck would find and collect "the best preserved, most complete example known" of *Pentanogmius evolutus.*[3] That specimen is now on display in the Denver Museum of Natural History. Complete except for its tail fins, it contains a great number of scales and the large sail-like dorsal fin, which is rarely preserved in other specimens of *Pentanogmius*. (Dad's *Pentanogmius* at the Fort Hays museum had a dorsal fin, but it was damaged and collapsed under the fish's body.)

When my parents did the bookings for the Plaza Theater, they chose a variety of offerings—from quiet independent films to lavish CinemaScope epics. Looking back over years of bookings, Pop said *High Noon,* starring Gary Cooper and Grace Kelly, was the best moneymaker for the investment. It was shot in black and white, and none of the other regional movie houses wanted to book it. It turned out to be a sleeper hit, and the Plaza Theater was the only venue that had booked it in western Kansas. Due to its popularity, Dad reserved it for an extra week. His overall rental cost was twenty dollars, and the movie generated a packed house for two full weeks.

Mom and Dad were booking movies long before the Motion Picture Association of America (MPAA) rating system of 1968, and some shows pushed cultural boundaries. The MPAA created G, PG, M, X (now NC-17) ratings, primarily to provide parents with an indication of a movie's content. Pop recalled how *Duel in the Sun* (1946) showed Jennifer Jones's torn blouse revealing too much skin, and the local Catholic priest discouraged his parishioners from attending. That did not stop Pop from booking it another week!

Blockbusters of the 1950s presented unusual settings and themes for western Kansas audiences, such as *On the Waterfront* (1954) featuring Marlon Brando; *Rebel Without a Cause* (1955) with James Dean; and *Jailhouse Rock* (1957) starring Elvis Presley. One of the most popular dramas was *Picnic* (1955), a film adaptation of a play by William Inge, the award-winning playwright from Independence, Kansas.

The late-1940s and 1950s were also the heyday of the big-budget Hollywood musical. *Carousel* (1956), *Oklahoma!* (1955), *The King and I* (1956), and *South Pacific* (1958) were just a few of the movies that captured the imaginations of Kansans and whose songs stayed in the Bonner playbook for years. *Oklahoma!*, in particular, resonated with rural audiences, with such melody lines as "Oh, the farmer and the cowboy should be friends," and "The wind comes sweepin' down the plain" striking a familiar note. In that movie, the plains dwellers were seeing a romanticized version of their own history.

Oklahoma! was a movie that also made me cringe. It symbolizes the era it captures, with set-in-stone gender roles, rivalry between cowboys and farmers, and the idealization of the white settlers with no mention of the Native Americans who were pushed out. The movie proudly and baldly displays white-centric situations that modern writers do not swallow wholesale.

From the silents to the Plaza's closure, Westerns reigned supreme, and their propaganda dominated our upbringing. Zane Grey (1872–1939) was a popular novelist who influenced the development of the western genre. His works idealized the Old West and helped create the archetypal western hero.

According to Grey's biographer, Steven May, "The ascending arc of Grey's career matched that of the motion picture industry. It eagerly adapted Western stories to the screen practically from its inception."[4] Zane Grey worked directly with Hollywood producers, and such films as *The Lone Star Ranger* and *West of the Pecos* were based on his novels. Between 1911 and 1996, from the silents through the talkies, 112 films were adapted from his novels and stories. The Lone Star Ranger character eventually morphed into the Lone Ranger, a saga that migrated to television.

A teenaged Orville Bonner enters the family's movie theater, 1953. The winds of western Kansas have weathered the marquee since its installation in 1941. The poster advertises the film Lili, *starring Leslie Caron.*

Grey's best-selling novel of all time was *Riders of the Purple Sage* (written in 1912), which was made into a movie many times. *Riders of the Purple Sage* was set in Mormon-governed Utah, but Grey took inspiration for his stories from many western locales, including Kansas. His second novel, *The Last of the Plainsmen* (written in 1908), was based on the life of Charles "Buffalo" Jones, one of the co-founders of Garden City.

In a nice turnabout, after Dad closed the Plaza Theater in 1960, he began reading aloud to us in the evenings, and the novels of Zane Grey were a staple. When Dad read *Riders of the Purple Sage*, Mom helped portray the drama. In one memorable line before a gunfight, Dad read, "The breaking of Nell's muscle-bound rigidity. ... Then he drew." At this moment, Mom, off to the side with her knitting, yelled, "Bang! Bang! Bang!" which made us all jump.

Chuck says of the picture show years, "I lucked out because I had the best of both worlds in the theater. I remember helping sweep out a few times, but I didn't have to do any of the hard work." Chuck was ten when the movie theater ceased operations. The last movie was *Thirteen Ghosts*, shown in the fall of 1960. The year the theater doors closed, Steve, a senior in high school, was essentially running it. Steve graduated from Wichita County High School in 1961.

The townspeople of Leoti were gravitating toward television, but the theater building itself was becoming dilapidated and the machinery was wearing out. Steve remembered that Dad did not fix or replace the projector because the income from the picture show was trickling away. Steve said that toward the end of the theater days, the uptake portion of the movie projector didn't work, so he had to rewind the reels by hand. At the time the Plaza closed, ticket prices were fifty cents for adults and twenty-five cents for children. When Steve was showing movies, as he put it, "for only a couple of loyal couples," it was obvious the business couldn't stay afloat. In addition to TV, there were competing youth activities such as Little League baseball, which gave Leotians an alternative to the movies.

Another contributor to the loss of revenue at the theater involved the seasonal Mexican migrant workers who came to Wichita County as farm laborers during the summer. The movie theater was one of their only entertainment options. There was no segregation in Kansas as there was in the South, and Dad didn't prohibit anyone from attending based on race. But sadly, many members of the white community stopped coming. During its early days, Leoti, like most western Kansas communities, was predominantly white. A few Latino families began settling in the town starting in the late 1920s, but very few Asian Americans or Black people have ever lived there. Currently, the number of Leoti citizens who identify as Hispanic or Latino comprise almost 40 percent of the population.[5]

Because Dad cited television as a main reason the theater stopped being profitable, he resisted having a TV set in the house for years. That resistance crumbled in 1968 when Clare Jane and her husband, Ray Askey, brought a black-and-white television to the Leoti house on one of their visits. It had rabbit ears, which might have worked in a city but not in a remote town like Leoti. To pull any station in, we soon realized we needed an outdoor antenna. So Steve bought and helped install an aerial that was positioned outside the roof on a three-quarter-inch galvanized pipe. The pipe ran up from the heater room to the roof, and to get reception, one of the boys turned or shook the pipe until the viewer on the other side of the wall yelled, "There! That's it!" Most of the shows were moderately snowy, but we could pull in the Garden City NBC affiliate, and sometimes, if we were lucky, we could pull in CBS from Goodland. That one took a lot of adjusting. Dana remembers: "After someone in the living room told me when the picture was coming in, I would clamp the pole so that it stayed in place." But with the raging winds of western Kansas buffeting the aerial, it wasn't long before the clamp jerked loose.

Not that my brothers and sisters really needed television. All of us could entertain ourselves and others quite well. Letty in particular was a "class clown," a jokester with many creative ideas. And of course, Dad entertained us while driving, roaming around the house, or collecting a fossil. The autodidact spouted verses such as Longfellow's "Listen my children and you shall hear / Of the midnight ride of Paul Revere." For Father's Day in 1960, Mom gave Dad a fifty-cent book with a pseudo-suede cover called *Popular Poems*. In the small book were favorites such as "Toledo Slim," "Casey at the Bat," "The Face on the Barroom Floor," and "Passing of the Back House." He quickly committed to memory the ones he liked best.

Despite the dismal downturn and closure of the theater, not having to spend time running it meant Mom and Dad could open themselves up to more school activities, and our family could see our brother Steve's and sister Chris's performances. The years from 1960 on would make possible a frenzy of fossil collecting as well.

Digging the Dirt
1950s

IN LATE 1948, WITH the winnings of a poker game combined with earnings from the theater, Dad purchased a quarter section of land south of Leoti. Two years later, he bought a second quarter section near the first one. He prosaically dubbed them Number One and Number Two. In addition to wheat farming, these parcels of land gave the family space to raise livestock: hogs, chickens, turkeys, and a few head of cattle.

The 1950s brought high income from wheat crops and the theater, making Dad and Mom, survivors of the Depression, prosperous for the first time. During this boom period in the motion picture industry, the theater brought in twelve hundred dollars a month. The show was doing so well that it prompted a wealthy landowner to offer to trade four quarters of land for it. My parents thought about it, but politely declined. They thoroughly relished their means of making a living. In cinema's heyday, no one could have predicted the eventual decline of small-town movie theaters. Moreover, Dad had enough experience with wheat farming to know that he didn't want to engage in that profession at a larger scale. As it was, Number One and Number Two consumed much of his spare time.

In "Wheat People," the Kansas Historical Society summarized the importance of the wheat harvest in western Kansas: "Harvest is the most important time of the year for rural residents. It means a paycheck for everyone—farmers, custom cutters, and local businesses. The local grain elevator becomes Grand Central Station during harvest. People, trucks, and wheat constantly move in and out."[1]

My two oldest brothers, Orville and Steve, worked extremely hard on these parcels of land, helping with wheat planting and harvesting. The main implements were a Caterpillar tractor and thresher, plus a farm truck. On Number One, they planted wheat on all but twenty acres that surrounded some outcroppings near Sand Creek. These were Tertiary Epoch bluffs,[2] a later geological period than the Cretaceous. Dad also planted wheat on Number Two and had an excellent yield on it in 1952.

He made enough money on the 1952 crop to take the family on a real vacation, touring the mountains around the Great Plains. He and Mom loaded up my six brothers and sisters in the big blue Olds sedan, and they went to the Rocky Mountains of Colorado, then to the Black Hills of South Dakota to see Mount Rushmore. Chuckie was two, so he sat on Mom's lap most of the time and crawled around on the other kids. Along the way, they visited Dad's cousins Cye and Harry Bonner in Lusk, Wyoming; the "Tower" in Genoa, Colorado (which Uncle Jennings would purchase later); Uncle Ralph's family and the natural history museum in Denver; and Dad's birthplace, Imperial, Nebraska. He showed the kids the house where he was born and the graves of Bonner ancestors, regaling them with stories and recounting the migration from Nebraska to Kansas.

When Dad first started working on the parcel of land known as Number One, he and his helpers cleared it to make farmland, removing an old house and barn. They put in a new fence and disced the soil for planting.

One day in the spring of 1949, Orv, Dad, Uncle Eldredge, and his son, Bob, were working on a fence on Number One. The new green wheat crop looked fresh and strong against the black topsoil. They all looked at the promising crop with pride. But they also noticed, off to the northwest, an ominous navy blue cloud. Our father had experienced the 1923 tornado in downtown Leoti as a boy of twelve and was convinced that if he could keep any storm within his sights on the vast plain, he would be able to evade it or outrun it. He had done so several times up to that day. "Boys, we gotta get in the car *now* and outrun this damn storm!" Dad yelled down the fence line. "Let's go!"

The adult brothers and the two cousins ran to the four-door sedan, a Ford sedan Pop called the Fan, whose doors opened in opposite directions. Dad sped off, driving seventy miles per hour along the dirt road. They headed south toward the Wichita-Kearney County line. After several miles, noticing the dark cloud getting closer, Dad started feeling a strong wind pushing the car and knew he couldn't outrun the storm this time. He stopped abruptly in the middle of the road and hollered, "Boys, hit the ditch!" All of them bailed out of the car. Dad and Orv plastered themselves down in the driver-side ditch, while Eldredge and Bob sprang into the ditch on the other side. All four doors of the vehicle stayed open, which may well have kept it on the road. The trunk blew off and went straight up into the air, circling and landing in the pasture beside them.

After the wind calmed, they went into the pasture, retrieved the trunk door, and returned to Number One. The late May wheat had been reduced to stubble. The wind had sheared the spikes off the stems, while the hail accompanying the storm pounded it into the ground. That crop was a bust—it would have been the first wheat crop for our family off that land.

Providing relative freedom was a hallmark of Mom and Dad's parenting style, but if a child was unruly, order needed to be maintained. When it came to discipline, Mom was hands off. She expected our father to handle it. If we made her frustrated, she chastised us with guilt. If extremely frustrated, she would cry, which set Dad off. Heaven help you if you provoked a tear from her eye. We were only physically disciplined if we broke a cardinal rule, such as lying, talking back disrespectfully, getting drunk, or missing curfew. The boys, particularly the two oldest ones, received more whippings than the girls.

Serious physical discipline, however, was rare. And with so many children running around, no one was monitored or "helicoptered." We were on our own. Our parents allowed us to grow within boundaries. Often this involved helping a child work through a problem. Case in point: In 1951, fourteen-year-old Orv decided he wanted to make some nesting coops for a turkey venture he was starting on Number One. His plan was to raise the chicks in the garage in Leoti; then when they got big enough, transfer them to the coop on the farm.

One frigid day, Orv decided to build the coops in the kitchen of the house in Leoti. The completed structure was so large—four by four by six feet—that it wouldn't fit through either the kitchen door or the door to the garage. So, Dad suggested a plan. Instead of having Orv dismantle the large unit, they removed the windows from the kitchen and lifted it out of the house that way. It seemed perfectly logical. The man who built the house saw nothing wrong with temporarily removing and reinstalling the windows. Orv laughs about this story to this day.

The turkey-raising story has a dreadful ending. The poultry shed on the farm was raised up off the ground and sat on cinder blocks. Badgers went underneath the floor of the shed, and even though they couldn't get inside, their presence frightened the young turkeys, who stomped themselves to death.

Orv's story about the coops illustrates how Dad and Mom allowed us kids to experiment with our own interests and find our own paths. After Orv got back from his service in Europe in 1956, he tried his hand at raising Landrace hogs. Orv's adventures with livestock allowed him to sample various career options, and for a time he lived out on Number One in a small trailer. But after he had experienced wheat farming and turkey- and hog-raising, he decided to try college. Because he was so involved with Dad's fossil hunting and collecting and had gotten to know George Sternberg, he began to think paleontology was a path to consider.

After the livestock experiments were over, the shed they called the chicken house took on a second life. The older boys disassembled it and brought it from the country into Leoti, where they placed it on cinder blocks on the second lot of the house near the alley. They transformed this shed into a playhouse for the younger kids. Steve added swinging saloon doors and Letty created a post-office area, with wooden soda crates for mailboxes. The chicken house was a magical place where we could swagger up to the bar and ask for a sarsaparilla, as well as request our mail from the postmaster. Another generation making our own entertainment.

We younger children also enjoyed a treehouse in the box elder tree north of the house. The boys in particular spent many hours up there, setting the dials of their transistor radios to KOMA, an Oklahoma City radio station that aired Top 40 hits. It was a favorite clubhouse escape, but it only safely held two teenage kids or three smaller ones at a time. I spent many an afternoon up there reading comic books. We were arboreal. (In later years I climbed every tallish structure I could find in Leoti: The water tower, the courthouse, and the Catholic Church were my favorites. It felt thrilling and gave me a different perspective on the town. I confessed these high school hijinks to Dad after I left town.)

The influence of the library on our family was also significant. Grandmother Berg served as the librarian for almost twenty-five years, from 1936 through 1960. The first library was in the Wichita County Courthouse, in between the Plaza Theater and our house. In 1958, a new one-story brick Wichita County Library, just south of the courthouse, opened its doors. From the moment we could walk, Mom took us to the library to pick out books.

Grandmother liked to spend time with one grandchild at a time during the school year, and my brothers and sisters visited her in the library often. Letty remembers helping her and learning the Dewey Decimal System. During the summer, Ralph's two older children, Pam and Bruce Berg, stayed in Leoti at her house. My memories of Grandmother are limited since I was so little when she passed. I do remember she gave me a bag of Planters peanuts every time I visited, and she allowed me to call her "Mimi" but expected the other kids to address her as Grandmother Berg. I also recall her little square house and visiting her in the hospital in Leoti before she died.

Mimi died in 1961 at the age of eighty, only one year after she retired from the library. We did not have a phone, but a tall man came to our door from the hospital, informing Mom of Grandmother Berg's death. (I found out later that the tall man was her doctor.) Her funeral was held in the First Presbyterian Church, of which she was one of the original members, and she is buried in the Leoti Cemetery. Her two children, Ralph and Margaret, gave her eleven grandchildren. And before her death, she knew four great-grandchildren, Clare Jane's and Letty's two oldest kids.

After leaving Leoti in 1935, our Uncle Ralph Berg moved to Chicago and lived with his Aunt Lillian, working in various jobs. After service in the Army, Ralph worked as a traveling salesman, like his father, and frequently came back to Leoti to visit his mother and sister. Sales was a perfect occupation for him; he was naturally loud and gregarious, the opposite of Mom. He lived most of his life in Chicago, Denver, and San Antonio, Texas. Ralph married Viola (Andrews) Berg in Chicago and they had three children: Pam (1947), Bruce (1948), and Penny (1959). He and Viola divorced in the late 1960s, and he later married Betty Stauffer, his former high school sweetheart from Scott City. Ralph died in 2002 and is buried at Fort Sam Houston National Cemetery in San Antonio.

Pam, our oldest Berg cousin, remembers visiting Leoti in the summertime, staying with Grandmother Berg. She went to the library with Grandmother during the day, then to the movies at the Plaza Theater almost every night. She has fond memories of Mom, saying, "Aunt Margaret would make sugar popcorn balls with the left-over popcorn, and she also made us animal-shaped pancakes." When it came time to go with the Bonner cousins to the fossil beds, Pam says, "As a city girl, the idea of going to the fossil beds was exciting. The reality, though, was it was hot and dusty. I didn't really know what we were supposed to be looking for, but we had fun playing out there."

To Dad, gambling with Leoti cronies was a fun way to blow off steam, but it served a larger purpose—adding to his income. His game was poker, and he liked to play stud and sometimes dealer's choice. He honed many poker expressions: "aces and eights—the dead man's hand" ... "deuces loose, down the river" ... "bullets for O'Hara" (upon seeing aces). Many of the men he played poker with had been in the war, and in an era when men didn't talk about their feelings, most of them started drinking heavily. Dad joined them in both cards and imbibing, and this heavy drinking became an issue for our family. Mom handled him well, and the children were usually in bed when he came home drunk. Clare Jane remembers that the sign to expect a partying Pop was if he didn't show up for dinner. Dad drank occasionally until the early 1960s, about the time the theater closed. We're not exactly sure why he stopped then. Mom might have suggested it, strongly. Her mother was gone, and she needed him at full capacity.

Steve remembered the actual poker hand Dad held when he won the largest pot of his life from another World War II veteran, a wealthy wheat farmer. "Dad was playing five-card stud, with one hole card, jokers wild." Steve related. "Jake Brown [not his real name] had two aces showing. Dad bet high because he also had two aces showing." Brown felt confident with his aces because he had two kings behind them. So with all four aces showing, Brown studied the cards, then called Dad's bet, saying, "Bonner, you'd better not have that Cuter in the hole." "Cuter" was their term for the joker. "But Dad *did* have the Cuter in the hole," Steve recalled. When the two men turned over their hole cards, it was three aces beating aces and kings.

With the winning pot of fifteen hundred dollars, Dad purchased the lumber to frame up the big house. It was the last time he played high-stakes poker. The next weekend, one of the regular players came by the house to entice Dad back into another poker night, aware of his big win the previous weekend. "I don't have any money to play. My money is sitting right there," he said, pointing to a pile of two-by-fours.

Dad had been around gambling all his life. In the Army he supplemented the family's income by playing poker with his and Mom's Army stipends. There is a line in Zane Grey's novel *The Lone Star Ranger,* where the main character tries to entice a man into a game of chance. He says, "You're a gambler, Poggin. You're the ace high, straight flush, hand of the Big Bend." Dad loved the bravado of this quote and probably subconsciously applied it to himself .

From the Arkansas days on, my parents brainstormed about building a large ranch home in Leoti with Prairie-style echoes and Thermopane windows. "Thermopane" was a brand name for insulated double-paned windows with an airtight seal between the panes. My older siblings remember the Thermopanes were a big investment, costing five hundred dollars. Mom and Dad wanted to build a small rectangular home first, which they did after the war, then add on an extension that would transform it into a long, bracket-shaped structure with a matching rectangle on the opposite end.

The larger home was finished in 1950. The ell on the original nine-hundred-square-foot structure expanded the home to twenty-five hundred square feet. A fashionable addition at the time was blue porcelain bathroom fixtures, something that impressed our Leoti friends for years. The Thermopane windows performed as planned. Facing south, the windows allowed the sun's rays to reach far into the living room in winter, but only a couple of feet in during the summer. The house didn't have central air, so controlling the sun's reach and using circulating fans kept it comfortable during summer months.

By the time Dad and Mom got to child number seven, they were running out of family names. Dana Andrews Bonner was named after the movie star Dana Andrews (Mom must have thought him a handsome actor!). "Andrews" was also the maiden name of Uncle Ralph's wife, Viola.

Clare Jane remembers that the only time the garage in the new end of the house was used for a car was in the days before Dana's birth in January of 1953. Dad, worried it might be too cold outside for the car to start, had never built a concrete ramp to the garage, and the entrance to it was curb-height. So he laid two sturdy doors down and drove the car up over the curb into the garage. The Blue Olds started right up when the time came to make the journey to the doctor in Scott City. After that, the garage served as storage and Dad's fossil workshop.

At play in the fossil beds, Chris, Chuck, Letty, and Dana plan a house layout using rocks and canyon junk for features in the rooms, 1956.

Like Clare Jane before her, Letty became responsible for looking after the younger siblings. In the fossil beds, Letty remembers, she was personally a little impatient when it came to fossil hunting. She preferred to play with Chris, Chuck, and Dana. She didn't see taking care of her brothers and sisters as a job.

Whether in the fossil beds or at home, Letty invented plays for us to perform, such as her comic "Perils of Pauline." An iconic photo of young Letty in the fossil beds shows her standing with one of Chuck's diapers serving as a scarf to hold down her hair. Indeed, the wind presented problems to us, particularly when our hair was long. We held it down with scarves and caps, but ponytails, pigtails, or braids were most effective.

When the family all jumped out of the car, whether in the fossil beds or on the farm, she would say, "Let's play!" Leadership of young children helped define Letty's future vocation. In 1969, when her own three children were small, she attained her bachelor's degree in education from Arizona State University in Tempe, Arizona. Two years later, she earned her master's degree in education from ASU as well. She had a thirty-five-year career as a first- and second-grade teacher in the Phoenix and Tucson areas.

Dana's first memory of the fossil beds is almost spiritual. He was old enough to traipse around by himself, probably about four, so he just wandered off one day. He looked at bugs, listened to the wind, and walked until he got to a fence, where he observed several cattle chewing their cuds and looking lazily at him. After a while, an inner voice said, "You'd better go back," so he turned around and headed back in the direction from whence he had come. Meanwhile, the rest of the family had been searching, shouting, and getting frantic. As if in answer to their calls, his little round head appeared over the hill.

One of Dana's frequently quoted comments was when Chuck, seven, excitedly showed him a shard of something gray and shiny, yelling excitedly, "I found a fossil! Look!" The four-year-old Dana's unimpressed response was: "Aw, Jackie [Chuckie], that's just gypsy [gypsum]." As they got older, they developed distinctive styles and expectations in the fossil beds. Dana was serious about hunting and happy to let others do the collecting. He found many quality specimens in his Kansas years.

The 1950s saw the older Bonner children maturing and leaving the nest. Orville graduated from high school in 1954 and served in the Army from 1956 through 1958. Clare Jane married her high school sweetheart, Alton Mitchell, in 1956 and graduated that same year. Letty graduated and married an Oklahoman, Buddy Bristow, in 1959. But Mom and Dad also added three more children in this decade: Chuck in 1950, Dana in 1953, and me in 1957.

Dad was forty-six and Mom was forty the year of my birth. A few months later, their first grandchild, Clare Jane's daughter, was born, so I became an aunt at three-and-a-half months. Clare Jane, excited to be pregnant, said to Mom, "I've got news! I'm going to have a baby!" Mom replied jokingly, "That's nothing—so am I!" As they prepared for the arrival of their children together, Clare Jane could not have had a better model of motherhood.

Neither of my parents ever got an annual physical to be sure all was well. Mom never went to the hospital unless to have her babies. She had all her children except one in the hospital in Scott City, twenty-six miles east of Leoti. The exception was Chris, who was born in Garden City. I can just picture Pop's fevered driving to these neighboring towns whenever Mom was ready to deliver.

When we started taking a more active role in fossil hunting, we four post-war children had different approaches. Chris was the micro hunter. She focused her efforts on finding tiny fossils. When I got older, I emulated the way Dad hunted. Chuck would zip off into the far ends of the draws, and Dana loved to explore the upper reaches of the bluffs, so they were named the Coyote and the Mountain Goat, respectively.

In the mid-1950s, Dad's discovery of a nearly complete short-necked plesiosaur changed the whole family's perspective toward fossils. This fossil was the best of its type in the world and bonded Dad with renowned paleontologist George Sternberg at the Fort Hays Museum of Natural History. The Bonner family's view of paleontology as an entertaining family pastime changed to a serious scientific vocation. Everyone wanted to find something important.

As the 1950s drew to a close and the picture show was flagging, Dad knew he needed to close its doors for good. As he had done during the Depression, he needed to scare up another means of income. Although he had had stellar wheat harvests in that decade, he was weary, and wary, of farming.

Western Kansas farmers were turning increasingly to center-pivot irrigation rather than being held hostage to the vagaries of weather, as dry land wheat farmers were. (Center-pivot irrigation creates those round circles of green you see when flying over the High Plains.) Dad didn't care to dig wells and make that kind of investment, and it also went against what he believed about the Ogallala Aquifer, the water supply that farmers were using to water their crops. He believed the aquifer could not be replenished at the rate it was being used, and referred to it as "fossil water." Geologists agree with that assessment. In *Kansas Geology,* Rex Buchanan wrote: "The huge supplies of ground water contained in Tertiary and Pleistocene deposits of western and southwestern Kansas are among the most important natural resources in the state. Like oil and gas, these supplies were once considered to be limitless, but hydrologists now know that water is being pumped out for irrigation and other uses faster than it is being naturally replenished."[3]

Dad was able to see the handwriting on the wall. In the modern era, the small family farms of Kansas have mostly been absorbed into larger farms, which use improved seeds, machinery, and more efficient farming methods. He sold Number One and Number Two in the early 1960s and banked the money to use for remodeling the theater building. He and Mom wanted to repurpose the large structure and lease it to a retailer. They believed the building's rent would provide enough income to keep their days free for "fossiling." Dad knew selling fossils would never be lucrative, and he also knew the Kansas museums were well stocked with fossils from the Cretaceous beds. Even so, he wanted to try to spend as much time as possible in the field.

At the 1960 Society of Vertebrate Paleontology meeting, which took place in Chicago, he met Dr. Robert H. Denison, the curator of fossil fishes in the Chicago Museum of Natural History, also known as the Field Museum.[4] The Field Museum is a glorious classical revival edifice strategically placed in the metropolitan Chicago area near Lake Michigan, and Dad first visited it with George Sternberg. From 1960 through 1970, Dad sold fossils to Denison on spec. In a letter to Dad at the beginning of their affiliation, Denison was almost apologetic about the amount the museum could pay him. The museum's authorized payment, he said, "should cover your expenses and a bit more, but I'm afraid you won't get rich collecting fishes at this rate. ... Many thanks for sending these specimens. I am certainly glad to know a collector near the Niobrara Chalk."

The Field Museum's first purchase, in 1960, was a *Squalicorax* shark that, at death, was extremely flexible. As Dad unearthed it, he marveled at the circular arrangement of the shark's vertebrae. The natural forces that caused its head to curl back upon its body may also be why the shark ended up nearly complete. He initially thought the curlicue shark was a chimaeroid, but the museum determined it was a *Squalicorax*.[5] A partial list of specimens that ended up in the Field Museum, in addition to the shark, includes *Elasmosaurus* material in 1960; a large *Pachyrhizodus*, two *Apsopelix*, and a *Saurodon* in 1961; a *Saurodon*, *Cimolichthys*, two *Clidastes* mosasaurs, and *Pteranodon* material in 1962; a *Protostega* and *Cimolichthys* in 1967, and a *Lophochelys* turtle in 1970.

The Niobrara Cretaceous turtle *Protostega* was an impressively large reptile, reaching a length of eight feet and resembling the modern leatherback turtle. That Dad sent some *Protostega* material to Chicago meant Denison was interested in reptile as well as fish material, adding this large turtle and other reptiles to their collections. Pop was proud to have formed this relationship with the Chicago scientist and happy the fossils were going to another prestigious museum.

For fossil hunters, nature provides many nuisances and dangers, such as rattlesnakes, heat, cacti, and biting insects. When Dad encountered rattlesnakes, he automatically killed them. To modern sensibilities, that reaction might seem extreme, since we were in their habitat, but his instinct was to protect his family. Because snakes were everywhere, he reasoned, random kills wouldn't upset the balance of nature.

In the field, Dad finishes putting plaster in the frame on a fossil shark collected in 1960. This specimen was shipped to the Field Museum in Chicago.

In a section about hunting fossils, the authors of *Roadside Kansas* warned: "Fossil hunters should remember a couple of things before they wander into the chalk bluffs. First, nearly all of this area is private property, and they should secure the owner's permission before they enter a pasture. Second, this area is home to rattlesnakes, which are active during the warmer months of the year."[6]

It's amazing that in all the years we went fossil hunting, no one experienced a rattlesnake bite. Dad knew of ranchers who had gotten bitten near fences, so we were always cautious around them. In the outcroppings, he instructed us to stay alert, tread slowly, and "keep your eyes peeled and your ears open" for the warning buzz of a rattle. He discouraged running because it wouldn't give us time to react if we encountered a snake. He said, "Reptiles are cold-blooded, so when it's hot out, they move fast. If the lizards are zipping, the snakes will be out too."

Even so, the competitive nature of the human hunters was sometimes hard to contain. One time Dana was racing down a ravine to get to the next one before Chuck did. He heard a sharp buzz and instinctively leaped into the air, effectively hurdling the snake, which was a thick, five-foot-long, pregnant female. We knew this because Dad followed behind Dana, killed the snake by chopping off its head with his Marsh pick, then sliced open the belly, which was full of live babies. We didn't know until that point that rattlesnakes bore live young, but that was just one example of learning about the natural world on the fly, or in Dana's case, on the jump.

Dana remembers an unnerving sight on the west side of Lone Butte. He observed a den of at least a dozen rattlers under a cliff. "I gave them a wide berth," he says. "They were having a rattlesnake party, draping around each other and buzzing in excitement." Lone Butte is a landmark in southwest Logan County that can be seen for miles in each direction. It was used as a marker by Native Americans, and later, cavalry soldiers and cowboys. We sometimes used it as a meeting spot for George Sternberg and others.

As Dad aged and we all went our separate ways, his attitude toward rattlesnakes mellowed. He respected their place in the natural world and generally left them alone, calling the big ones "old residenters."

We saw lots of nonvenomous snakes but ignored them, concentrating on catching the other plentiful reptiles: lizards. Once we understood to grab them by their bodies instead of their tails (which broke off so they could get away) we also learned other tricks, like putting them on their backs and stroking their bellies so they would play dead. The boys enjoyed letting them hang by one claw from a fossil tool. Their brown coloring was like that of rattlesnakes except the male lizards had an iridescent blue streak along their sides.

On one summer day, Dana, Chuck, and I each caught a lizard. Dana and Chuck named their new pets "Cope" and "Marsh," of course. I named mine "Lizzy the Lizard." We tried keeping them in our rooms in homemade terrariums. We experimented with giving them food and water, but they languished. Finally, we released them to run around Mom's flower garden outside the kitchen end of the house. Remarkably, they perked up and scurried away through the buffalo grass. They were likely prairie lizards or common lesser earless lizards.

Mom transplanted a few prairie wildflowers, which brought a touch of color to her garden in the spring and summer. The pastures around the chalk beds supported milkweed (with incongruous orange flowers), lavender-leaf sundrops, soapweed, sandwort, chalk lilies, and the yellow prairie coneflower. In the plants and the soil, modern nature seemed connected to prehistoric life. This description of the shortgrass soils of eastern Colorado and western Kansas reflects on the area's ancient antecedents:

> One hundred million years ago this area was an inland sea which resulted in the disposition of sand and shelled invertebrates creating the underlying bedrock of sandstone and limestone. [70] to 80 million years ago the Laramide orogeny of mountain building created the progenitors of the Rocky Mountains and subsequent periods of glaciation and erosion created huge outwash debris which formed the overlying deep soils of the present day.[7]

After working the deep soils of Number One and Number Two and selling the theater, Pop's days as a dryland wheat farmer and a movie theater operator were over. He became a landlord for a few years, then sold the remodeled theater building in the early 1970s. For the rest of his life, his focus was fossils and family.

The post-war years through the late 1950s were the most financially successful ones Mom and Dad would ever see. The older brothers and sisters received a wholly different upbringing from the younger ones, but only materially, and only momentarily. The emotional and intellectual support of our parents remained steadfast through the years, whatever their income.

PART III - MUSEUMS

Previous page: A nearly complete *Hesperornis regalis* specimen that Dad discovered and collected in 1958 is shown in a diving position on display in the Sternberg Museum of Natural History, Hays, Kansas.

Grand Man
1950s-60s

IN 1925, POP STARTED hunting fossils under the guidance of Arthur Wedel. Thirty years after that, in 1955, he found his first major fossil and took it to George Sternberg. As Dad became more serious, he refined his skills as a collector and preparator with advice from George. It's highly likely he picked up some Sternberg vocabulary and musings while hunting and working on specimens. Dad tended to absorb things like a sponge.

George Sternberg, after Mr. Wedel, was the second scientist to guide my father in the Kansas chalk. He was the most well-known fossil hunter to grace our lives. His contributions to science are significant, and the full story of his family's and his involvement in paleontology appears in the next chapter. But beyond the scientific background, George was a stellar human being who had a positive influence on Dad but also the rest of the Bonners. The particulars about his personal relationship with our family deserve mention. Our memories of this genial man are interwoven with memories of our father.

Dad first met George in 1927 when he went to the museum in Hays to check out a *Xiphactinus,* the largest fish of the Cretaceous ocean. He met him again when George was collecting another *Xiphactinus* near Oakley, in northeast Logan County. George had become acquainted with the teachers in the small town when he lived there before moving to Hays, so when he uncovered another perfect, complete *Xiphactinus* in 1930, he invited Oakley's science students to come watch him collect the immense fossil fish. Sternberg always enjoyed teaching children: Before assuming his role as museum paleontologist at Hays, he "wished he could qualify as a teacher so he could work with children on a day-by-day basis, cluing them in to the wonders hidden in the rocks and wastelands. But with only a fifth-grade education, he knew his teaching would have to be out in the field rather than in the classroom."[1]

Pop heard about George's on-site demonstration and went to watch, enthralled, as Sternberg meticulously dug around the fish, preparing it for collection in plaster encased in a wood frame. For fieldwork, George wore khaki pants and shirt, with a fedora hat. While working around the fossil, he lay down on the chalk quarry and eyeballed the fish at ground level, spending much of the digging time in a prone position.

At times during George's excavation of the fish, Dad scoured the chalk gullies around the site and found a few shark's teeth. Dad secretly wanted to find something significant to impress the famous paleontologist, but it was not to happen, this time.

From watching George in later years, Dad soaked up his lingo and acquired practical collecting and preparing tips from his mentor. He used the word "scattered" when assessing a specimen's quality. The term meant the bones were "disarticulated," as if stirred up. When a fossil fish or reptile is articulated, it looks much the way it did in life—straight vertebral column, all the other parts connected where they should be. While unearthing a fossil, Dad would often proclaim if bones were scattered or articulated.

He also learned from George that field paleontologists choose the method of collecting based on the specimen and the quality of the material it is in. Sometimes the chalk itself is so hard that it is possible to simply carve the matrix and pry the chalk block with great care from the surrounding bluff. This is infrequent, however. While excavating fossils, Pop made appraising comments, such as "What beautiful material this is. Look how waxy this chalk is!" or "This is pretty cruddy stuff. God all Friday, it's flaky. I'm gonna have to be careful." The bones themselves could peel apart during fieldwork, and sometimes modern plants sought out the minerals in the bones. When that happened, the roots ruined the specimen.

If a fossil was articulated and in good chalk, Dad usually used the Sternberg "frame and pour" method of collecting, protecting the fossil in plaster in a rectangular frame. When a fossil was in poor, crumbling material, the plaster jacket was the best means of collection. For this method, Pop carved around the matrix of the fossil and then covered the rounded area with wet paper towels or newspapers. Over that layer, he sealed the chalk or shale holding the fossil with strips of burlap soaked in plaster, much like a doctor putting a cast on a broken limb. After the burlap and plaster hardened, Dad carefully scraped the rock away from underneath the fossil until he could safely turn the whole thing over. The resulting cradle held and protected the fossil.

Unrelated fossils—random pieces eroded from the chalk of bone, teeth, fins, shells from various species—were often picked up to keep. These separate fossils are not connected to a larger specimen but were parts of skeletons found at different strata and locations. As we were growing up, we Bonner children would fill up boxes or trays in our rooms containing these "keepers." In fossil-fishing terms, keepers are the ones that got away. Mike Everhart confirmed that in the Sternberg Museum storage area there are thousands of keeper fossils that George picked up over the years.

When George and Dad's friendship began growing in the 1950s, we did not have a telephone. The two friends wrote many letters back and forth, and our family donated the Sternberg side of the correspondence, which Dad had saved, to the Sternberg Memorial Museum in Hays. Their deep connection is clear in the tone of these letters, and in the increasingly personal discussion of family issues and other topics besides fossils.

"When are you coming over, Bonner?" was the question George posed in the letters and later on the phone. Dad always felt good when he heard this; to him it meant George valued his work and wanted to know what specimens he had found. But it also meant he wanted to spend time with Dad, who had become like a son. By extension, we Bonner kids became like grandchildren.

On more than one occasion, George brought his brothers, Levi and Charlie, to Leoti for a visit. Levi and Charlie Sternberg both turned out to be professional paleontologists in Canada. Of the three Sternberg sons, Charles M. was the one to bridge freelance fossil hunting and academia. He received honorary degrees from the University of Alberta, Calgary, and Carleton University in Ottawa, Ontario. He enjoyed writing scientific descriptions of his many fossil finds and was an honorary member of the Society of Vertebrate Paleontology.[2]

Almost every time the brothers visited George in Hays, he would bring them to our house. Charlie was the tallest of the three and, to me, the most serious. I remember Levi as an entertaining, impish character. He was even more diminutive than George, like a fun-loving leprechaun. He taught us tricks and poems. I was entranced and amused. As a small child, I had no idea that this playful man had discovered a nearly complete duck-billed dinosaur, *Parasaurolophus,* and collected large dinosaurs like *Triceratops.* Levi worked for the Royal Ontario Museum in Toronto from 1919 through his retirement in 1962. His colleague, Dr. Loris Russell, wrote of Levi: "In the field he was a hard worker and a congenial companion, always joking and teasing. He always enjoyed excellent relationships with local people wherever he was working. Levi will be remembered as one of the last of the old-time bone-hunters."[3]

George and Anna, his wife, were in the habit of making "field trips" from Hays to Leoti, and they sometimes stayed overnight. When we kids hit the hills and spread out among the ravines in search of bones, George was no doubt reminded of the good old days when he and his brothers worked in his father's fossil-collecting enterprise.

When our parents hosted fossil-hunting guests in the field, Mom made splendid lunches of fried chicken, carrot and celery sticks, fruit, cookies, and other delights. The Sternberg brothers raved over these picnic lunches and told us about their hunts in their early days of fossil hunting, when they ranged far afield from their base camps, ran low on provisions, subsisted on potatoes, and endured all kinds of deprivation to find the next big fossil.

Dad and George Sternberg in New York for the 1956 meeting of the Society of Vertebrate Paleontology.

George told us a story of driving his father's mule-drawn wagon from the outcroppings in southwest Gove County to Oakley in northwest Logan County to get food and supplies. Coming back, he was proud of the fact he was making good time, bringing all the items on the list so the hunters could survive a few more weeks. He remembered boards, plaster, and burlap sacks for collecting, but forgot one key item: tobacco. His father was not happy with this omission, so George had to turn the mules around and go back to Oakley for tobacco.

Both Dad and Orv told anecdotes about George. Orv remembers a time when he was a graduate student at Fort Hays and was invited to go fossil hunting with George and his brother Charlie. The hunting party was out in the middle of Gove County, and they were getting ready to go back to the museum in Hays. George and Charlie were always rivals. Charlie stated, "It's 112 miles back to the museum." George replied, flatly, "No, it's 115," and they bet a steak dinner on it. As they were nearing Hays, George took a small detour, and the total distance to the museum on campus was exactly 115 miles.

George was a nonconformist, so it seemed only natural that his wife also marched to a different drum. On their Society of Vertebrate Paleontology trips together, Mom and Dad noticed that Anna would do unusual things. One time she opened her suitcase in front of them and had loaded up half of it with cashews. She also sometimes wore her galoshes the entire day. In the time we knew her, however, we Bonner children, especially the girls, venerated Anna as much as we did George, and her individualistic qualities entertained us.

Dad ribbed George—gently, though, because of his deep respect for the man. Both were die-hard kidders. George pranked Dad during one of their trips to an SVP meeting when they were all riding together on the train. Dad and George could both grab a nap, anytime, anywhere. At one point in the journey, Dad was dozing and snoring contentedly. His head was thrown back, his face slack, and his lower lip puffed out at each exhalation. For the entertainment of the ladies, George carefully placed a short pencil along Dad's lower lip. Mom and Anna stared, trying not to laugh. They all wondered how long the pencil would hang there. Dad's snoring eased a bit, but he continued sleeping, pencil in place, for several mirth-filled miles.

My parents booked a romantic comedy for the motion picture show called *The Voice of the Turtle* (starring Ronald Reagan, of all people) back in 1947. It didn't really matter what the show was about or that the quotation from the Bible refers to turtledoves, because from that time on, Dad appropriated the phrase to refer to marine sea creatures. He knew there were turtles lurking in the fossil beds waiting to be discovered and would say, "Maybe we'll hear 'the voice of the turtle' today!" at the start of a hunt.

While Steve looks on, Dad and George Sternberg inspect a fossil turtle in 1956.

The voice was making some noise on one summer day in 1956, right after Dad and George had become friends over a *Dolichorhynchops* plesiosaur (details about that find will come in the next chapter). A family photo shows Steve watching Dad and George work in a turtle quarry, so I thought Steve found the specimen. But Steve clarified, "That was not my turtle. I am nearly sure Dad found it. Of course, as the photo proves, I was there when it was collected." Why was George collecting the turtle in the photo, I wondered? Steve explained:

> The bones didn't look familiar to Dad, and we took a few to Hays for them to help him identify the specimen. Myrl Walker was at the museum when we got there, and he and Dad thought it may be the smaller bones of a *Pteranodon*, but Dad wasn't sure, as it was foreign to him. George Sternberg came in and asked, "What ya got, Bonner?" He took a quick look and said "*Toxochelys* turtle." Then he climbed up a ladder, pulled down one of the Cope or Leidy books, turned a few pages, and pointed to the fossil turtle bones. As a young boy, I was unbelievably impressed. We collected it about a week later.

For this fossil, now in the Sternberg Museum of Natural History, Dad and George collected a total of one hundred bones and bone fragments, and it was the most complete *Toxochelys* to that date.

Toxochelys, an extinct sea turtle, is the most commonly found turtle in the Kansas chalk. Its carapace could attain a length of two feet. Its behavior was probably similar to that of modern sea turtles, laying its eggs on the shoreline of the Kansas Cretaceous sea.

As I look back, I realize that I never knew a grandfather. My own Dad was a fairly old father, and he was as indulgent to me as a grandpa might be, but my natural grandfathers were long gone by the time I was growing up. Mom's dad, the Berg patriarch, had died in the flu pandemic in 1918 when she was a toddler, and Dad's father, Pappy, died in 1951. The first five children in my family remember Pappy, but Chuck, Dana, and I missed out on him. George Sternberg was the closest thing to a grandfather that I knew during my childhood. Looking through the Bonner family photo albums of the 1960s, I am struck by how many times we documented George and Anna and then later George the widower when he visited our house in Leoti or we met up with him in the fossil beds.

Toward the end of his life, George entered Hillcrest Manor, a nursing home in Hays. Pop visited him frequently, usually making the trip to Hays and back during the day when we were in school. We were always eager to hear the stories

Dad brought back after these visits. We loved George and felt sad that his health was declining. Dad told us that George habitually reminisced about the horses, mules, and wagons from the early fossil hunting days. By 1967, he got in the habit of straying off. Dad was convinced that George was reliving his youth and that he wanted nothing more than to get to the fossil beds. After one such episode of wandering, George told Dad after he returned to the nursing home, "That damn team of horses nearly ran over me!"

The older I get, the more difficult it is to remember specific details about George and about those days. I remember generalities: a kindly gentleman, a self-taught scholar, a droll wit, and a caring embrace each time we departed from him. George Sternberg is a warm presence over my childhood.

The grand man simply became a part of our family, more so as the years went on and he was alone. The evidence I remember for this is he adopted the nicknames Pop gave us.

He called me "Mellie."

Members of the Bonner family in Leoti with George Sternberg in 1965 after he became a widower. Left to right: Dad, me, George, Orv, Chuck, and Dana.

Rare Friendship
1880s–1960s

GEORGE FRYER STERNBERG WAS born August 26, 1883, in Lawrence, Kansas. He and his brothers were trained in fossil hunting and collecting by their famous paleontologist father, Charles Hazelius Sternberg. The father and three sons would become the world's first family-run fossil-hunting enterprise.[1] All three Sternberg sons were born in Lawrence. George and his two brothers, Charles Mortram (Charlie), born in 1885, and Levi, born in 1894, all became recognized museum paleontologists, George in Kansas, and Charlie and Levi in Ontario, Canada.

Their father, Charles H. Sternberg, was born near Cooperstown, New York, and moved to Ellsworth County, Kansas, at the age of seventeen. He learned about the Kansas chalk in the 1880s when he visited his older brother, Dr. George Miller Sternberg, then stationed at Fort Ellsworth. The older Sternberg brother was among the first to discover fossils in the western Kansas badlands, and alerted O. C. Marsh, Joseph Leidy, and other paleontologists to the presence of fossils in the beds.[2] Dr. George M. Sternberg discovered and collected the type specimen of the "giant predatory fish *Xiphactinus audax*" in 1867–1868, and it was described by Leidy in 1870.[3] Dr. Sternberg was in Kansas as a military surgeon assigned to Fort Harker (established in 1866 as Fort Ellsworth) in central Kansas. He would later become the Surgeon General of the U.S. Army during the Spanish-American War.

The doctor's younger brother, Charles, took to fossil hunting as a career and became well-known around the country for the collecting of dinosaurs in the mountainous west and marine animals from the Kansas chalk. He described the lure and toil of his vocation in his 1909 biography, *The Life of a Fossil Hunter*:

> [A]lthough my struggle for a livelihood has been hard, often, in-
> deed, bitter, I have always been financially better off as a collector
> than when I have wasted, speaking from the point of view of
> science, some of the most precious days of my life attempting to

make money by farming or in some other business, so that I might live at home and avoid the hardships and exposures of camp life.[4]

After being exposed to the fossils around Fort Ellsworth, Charles studied paleontology at Kansas State University[5] in Manhattan under Benjamin F. Mudge. Although he didn't earn a degree from Kansas State, Charles offered to work in the field for Edward Drinker Cope, and later, became an employee of Othniel Charles Marsh. Therefore, he gained collecting experience from the two scientists behind the Bone Wars.

At the age of nine, George F. Sternberg went with his mother and younger brother Charlie to a site in Logan County near Elkader, where his father was working. It had been a hard season for their father, and he was feeling pressure to find some quality fossils to carry the family through the winter financially. George proved his worth in the chalk beds, where he made his first major discovery—a Cretaceous plesiosaur. George wrote in later years:

> I have a very vivid impression of coming suddenly upon several large pieces of bone washed down the side of a steep exposure. ... I remember running back to my father shouting, "I have found a fossil, I have found a fossil." ... In due time father went over with me to see what I had. I shall never forget the change of expression which came over his face when we arrived at the spot. Nor will I ever forget his first words, "George, thank God, you have found a plesiosaur."[6]

While his father was working on this significant fossil, George went to a flat yellow bluff nearby and carved his initials, age, and year into the chalk. In 1968, my brother Dana discovered this evidence of young George's presence in the fossil beds, deeply etched and weathered. Dad, Chuck, and Dana collected the slab bearing George's initials and gave it to the museum in Hays.

Since George was such a natural hunter, he started working with his father full time after the fifth grade. A few years later, when he was seventeen, he found another short-necked plesiosaur. Charles sold that nearly complete specimen to Samuel L. Williston at the University of Kansas, and it became the type specimen for *Dolichorhynchops osborni*. Williston named it for his mentor, Henry Fairfield Osborn. He wrote of the plesiosaur, "The specimen of *Dolichorhynchops osborni*, herewith described and illustrated ... was discovered by Mr. George Sternberg, in the summer of 1900, and skilfully [*sic*] collected by his father, the veteran collector of fossil vertebrates."[7]

George Fryer Sternberg carved his initials in the hard yellow chalk near the site of his first major discovery. Courtesy of the Sternberg Museum.

George frequently stated how much he appreciated his early training in paleontology, believing that learning at his father's side made more impact on his life than his schooling had. George's career held many highlights, starting from the days doing hard labor in the badlands and mountains of the American West. But he considered the top three discoveries of his career to be a "mummy dinosaur" found in Wyoming; a strange dome-headed dinosaur called *Stegoceras*, now in the University of Alberta Museum; and his fish-within-a fish, collected in 1952 from the Niobrara chalk of Gove County.[8] That fossil received its name because inside the prehistoric skeleton of a fourteen-foot *Xiphactinus* is a complete six-foot *Gillicus*. This remarkable specimen was one of the biggest attractions of the museum at Hays, where George worked from 1927 until his retirement in 1961. The fish-within-a-fish garnered national publicity. It was featured in a two-page spread in the July 19, 1954, issue of *Life* magazine.[9]

The Sternberg family—Charles and his three sons—collected large dinosaurs in the mountains of Canada, Wyoming, Montana, and other states. The immense beasts, such as *Triceratops, Hadrosaurus,* and *Albertosaurus*, were exceedingly difficult to collect because the weight of their bones sometimes broke the collection wagons. They had to be shipped to museums in heavy boxes that were loaded onto trains. It was extremely arduous work in difficult weather and inhospitable environments.

After these experiences, George came back to western Kansas. He chose to stay on the plains for a reason. He often told visitors to the Fort Hays State Museum about the hardships of collecting the large dinosaurs. "I much prefer the chalk beds of Kansas," he said. "The marine fossils intrigue me, and I think the beds will yield many exciting specimens in the future."

In an article for *The Aerend: A Kansas Quarterly,* published by the faculty at Fort Hays State University, George described the Western Interior Seaway: "During Cretaceous times a great shallow sea stretched across the Middle West, reaching from the Gulf of Mexico to the far north. It was bordered on the west by the ever-rising Rocky Mountains. Large low-lying land areas were filled with rank vegetation. Marshes and lagoons were to be found. These were the haunts of the dinosaurs living at that time."[10] With mountains being formed on each side, the placement of the Western Interior Seaway roughly predicted the future outline of the Great Plains.

Later in the essay, George reflected on the marvelous way the animals of the Cretaceous Sea turned into fossils. He explained why the specimens were so well preserved: "The fact that these animals lived in the water, died, and sank to the bottom of the ocean, there to lie in the ever accumulating muds until the flesh had decayed and the bones were buried safely concealed from the air, until the organic material had been entirely replaced by the various minerals carried in solution, was primarily responsible for their preservation."[11] This description for the non-scientist reader was similar to the explanation of fossilization Dad had learned from Mr. Wedel.

An event in November of 1955 gave Dad's paleontological career a serious turn. One Saturday morning, he was scanning a ravine of gray sediments below the yellow chalk in a Logan County site he called Hell's Half Acre, also known as Goblin's Hollow. Along the edges were washed-out fossil bones that looked something like mosasaur paddles but were shaped differently. Dad carefully checked the gully and found the spot where the bones were leading back into the bluff.

The entire family had come on the hunt, and my older brothers and sisters were tramping around, each on a mission of discovery. The three youngest children, Chris, Chuck, and Dana, still a toddler, started coming toward Dad, with Mom trailing them. Dad was examining the bones intently, tracing down the wash to where he might find other parts that had eroded out of the hillside. They were difficult to see because they were close in color to the gray chalk.

"Get those kids out of here!" he shouted at Mom. She froze and gaped at him. He rarely yelled in such a loud, agitated way. She immediately got the two small boys in hand and motioned Chris away as well. Later, after Dad had located all the exposed pieces of the reptile, she said, teasingly, "You disowned us over that fossil!"

"I sure as hell did!" he replied, grinning. "It's a once-in-a-lifetime find. I think it's the most important fossil I've ever found. I'd like to hear what George Sternberg at the Hays museum thinks of it."

Dad's 1955 discovery was a nearly complete short-necked plesiosaur, which he and Orv, then nineteen, collected and worked on during the winter. They took part of it to Hays to show Sternberg, whose brown hair was now completely white. His hair and wire-rimmed glasses gave him a scholarly demeanor. Dad donated the specimen, initially called *Trinacromerum osborni* but later modified to *Dolichorhynchops osborni,* to the museum. It is one of the best of its genus in the world. (A photo of the mounted specimen appears on page 53.) *Dolichorhynchops* ("long-nosed face") measured from ten to sixteen feet long. It ate small fish and possibly cephalopods, such as squids.

My second oldest brother, Steve, twelve at the time, remembered how "that plesiosaur took over our lives. It was handled very carefully and was placed on the entire dining room table when Dad, Orv, and sometimes Mom were working on the bones." By the time Dad took the trays of bones and the weird, flattened skull to Hays, the bones were painstakingly worked out of the rock. This meant George didn't have to remove the matrix from the bones; his main task was piecing the puzzle together and mounting it, not a small feat. He was extremely grateful for the care Dad took with the fossil.

There was another reason George was excited about the plesiosaur. His first important fossil at the age of nine was a short-necked plesiosaur, and his discovery in 1900 of an almost complete skeleton became the type specimen of *Dolichorhynchops osborni.* Williston described the significance of George's plesiosaur specimen: "Thirty-two species and fifteen genera have been described from the United States, and in not a single instance has there been even a considerable part of the skeleton made known." Quoting Williston's assessment in *Life of a Fossil Hunter,* Charles H. Sternberg added, "I am glad that the University of Kansas owns this splendid denizen of her ancient Cretaceous sea."[12]

Imagine George's delight when, fifty-five years later, Dad brought him an even more complete specimen of the same kind of plesiosaur, and Dad's had a skull. The only missing portion was the left front paddle.

Ironically enough, the scientist who had prepared and mounted George's plesiosaur at KU in the early 1900s was Handel T. Martin, the man who had declined Dad's fish skull in 1929 and asserted that the Smoky Hill Chalk beds had been hunted out. Williston wrote of Martin's work on the KU plesiosaur, "The task of removing and mounting the bones has required the labor of Mr. H. T. Martin the larger part of a year, and is as finally mounted, an example of great labor and skill on his part."[13]

George Sternberg himself would find out how extensive the labor of mounting so many bones was—it took him two years, between all his other duties, to mount Dad's plesiosaur.[14]

Not long after Pop found the plesiosaur, George and Myrl V. Walker, his colleague at the Hays college, co-wrote "Report on a Plesiosaur Skeleton from Western Kansas," which stated: "This very fortunate discovery was made by Mr. M. C. Bonner of Leoti, Kansas, and he has kindly presented it to the museum. The specimen was collected from a small exposure of Niobrara Cretaceous a few miles southwest of Russell Springs, Kansas. The nearly complete skeleton,

An in situ photograph of the plesiosaur found in 1955 shows the pebbly chalk and includes a nearly two-foot-long Marsh pick for size reference.

although somewhat disarticulated, was imbedded in the bluish chalky-shale member which lies below the reddish-buff chalk in this area."[15]

Left to right: George Sternberg, Orville Bonner, Fort Hays museum assistant Leford Wendell, and Marion Bonner with the plesiosaur during Sternberg's preparation, 1956.

A type specimen is the first physical example of an organism referenced when the species is formally described and named. Some animals have been typed by only their teeth (such as sharks) or by whatever part had fossilized, such as the gladius in squids. In late 1955, Dad brought an unusual fossil to George. It very much resembled a round flyswatter and was twenty-two inches long.

If a fossil is thought to be new to science, it is placed in the hands of trained researchers, those who can, with comparative analysis, determine if a particular species has been found before. If the researcher posits it is indeed new, it then becomes a type specimen and is the one against which successive fossils are compared. George initially called Dad's squid a "flyswatter." He sent photos to Dr. Bobb Schaeffer at the American Museum of Natural History for help determining what the fossil was.[16] In a letter dated January 1, 1956, George told Dad that Schaeffer said "the paddle-shaped specimen is the 'pen' of a squid and this form is called *Tusoteuthis longis* [*sic*], also that it's found in both the Niobrara and Pierre formations." Later, on October 19, George wrote, "Halsey Miller of the State Geological Survey will be here

Friday to pick up a few fossils he will try to describe, including your *Tusoteuthis longus* [sic]. Says he will name it for you if it proves to be new. Yours sincerely, G.F.S." The prospect of having a species named for him thrilled Dad.

The squids, or cuttlefish, of the Cretaceous period were a large part of the marine ecosystem and a favorite meal for the wide variety of predators living there—sharks, mosasaurs, and large bony fish such as *Xiphactinus*. There is fossil evidence of a squid inside the gut of a smaller fish, *Cimolichthys*, which probably choked on the squid, leading to its death.[17] Unfortunately, because the only thing that fossilized on these medium- to giant-size squids was a wide, expanded area (rachis) and the long central shaft (gladius), paleontologists can only extrapolate the full body structure of these creatures.[18] The most common species of squid found in the Western Kansas chalk is *Tusoteuthis longa*. *Tusoteuthis* ("crushed squid") preyed on other cephalopods, fish, and possibly even small reptiles.

Type specimens are not set in stone. Halsey W. Miller, of the Kansas Geological Survey, identified Dad's squid as *Niobrarateuthis bonneri* in a 1957 paper.[19] In the years that followed, Miller named two more squid genuses: *Kansasteuthis* and *Enchoteuthis*. More recent studies view these different squid taxa as one type: *Tusoteuthis longa*, the very identification that Dr. Bobb Shaeffer suggested to George Sternberg.

A paper presenting this argument was published three decades later, in 1987.[20] Its authors, Elizabeth Nicholls and Henry Isaak, said the differences in squid species were deformations caused by crushing during the process of fossilization and that it was misleading to identify them as separate animals based solely on the only parts that fossilized, the gladius and rachis. The current name, *Tusoteuthis longa*, which harks back to the earliest name used (Logan 1898) is now generally accepted. By arguing that previous teuthid species from the Niobrara Formation were all assignable to the taxon *Tusoteuthis longa*, Nicholls and Isaak effectively ruled out *Niobrarateuthis bonneri* (which has an oval rachis) and *Enchoteuthis melanae* (which has a pointed rachis). Both of these squid fossils, found by Bonners, were described by Halsey W. Miller as a separate genus and species.[21]

Pop instructed us in the names of Cretaceous animals, and like him, we used genus names for all of them with one exception: We called all squid material *bonneri*. (A photograph of the "flyswatter" fossil appears on page 160.)

Practitioners of paleontology now come in many shapes, sizes, races, and genders. In Dad's day, the field was dominated by white men, but that situation has evolved. In 2014, Lance Grande, curator emeritus of paleontology at Chicago's Field Museum, surveyed the large U.S. museums and determined that at the Smithsonian National Museum, 30 percent of curators were women, and at both the American Museum of Natural History and the Field Museum, 20 percent were women.[22] While those proportions in the early twenty-first century weren't as high as in many other scientific professions, they were a huge advance from 1956, when Dad attended his first Society of Vertebrate Paleontology meeting in New York and met Tilly Edinger, the only female scientist in the room.

Dr. Johanna "Tilly" Edinger (1897–1967) was a native of Germany whose main specialty was brain development. She worked initially at Frankfurt University's Geological Institute but, fleeing Nazi persecution, moved to London in 1939, then to the U.S. in 1940. She joined Harvard University's Museum of Comparative Zoology and initiated the field of paleoneurology. She wrote *Fossil Brains* (1929) and *The Evolution of the Horse Brain* (1948) and would become the first female president of the SVP in 1963.

Dad was impressed. He used Edinger as an example for how his daughters could pursue the field of paleontology if they wished. When he noticed that my sister Chris, a blond-haired tomboy, liked to climb the highest knolls, he said, "You know, if you look in the ant dens up there, you will see the ants collect tiny shark and fish teeth [micro fossils]." Chris was fascinated and started trying to find the tiny teeth. Dad added, "Paleontology doesn't have to be for boys. If you wanted to do more fossil hunting and collecting, you could train to be a scientist like Tilly Edinger."

Pop was beginning to understand the layers of career choices in paleontology. There are fossil hunters and collectors (field paleontologists), museum paleontologists (preparators), and scholars with doctoral degrees (scientific researchers, using basic and applied research), who work together to document findings and describe new species. In institutions of higher learning, there is a blend between research curators and university professors.[23] After Dad saw how Orville and George worked together on the plesiosaur, he was also seeing in his mind's eye a career in paleontology for his oldest son.

But that would have to wait. During the trips back and forth to Hays, Orv had received his letter of "Greetings" from Uncle Sam. The Korean War had technically ended in 1953, but the country kept the Selective Service in place for several years after that due to pressures from the Cold War. Orv went into the Army at the age of nineteen in February 1956 (having chosen, craftily, "Korea" as his top choice, he was sent to France), and was discharged in February of 1958 at the age of twenty-one. After Orv left for the Army, Dad designated my next oldest brother, Steve, as his running buddy, and they went fossil hunting together regularly.

Dad aspired to be a species of paleontologist similar to George Sternberg—largely self-taught and a field paleontologist at heart. But George, by virtue of his position at the Fort Hays State Museum, expanded his titles beyond field paleontologist, or collector, to include museum acquisitions, curator, and preparator. By observing the process behind the squid specimen, Dad saw how a single fossil could add to the breadth of scientific knowledge.

George was twenty-eight years older than Dad. Both men had a saucy sense of humor, and both enjoyed being in the fossil beds more than any other place. As the years went by, what started out as a professional relationship turned into deep caring for one another. Both men valued fossils that were rare, and both would concur their friendship was rare.

Understanding that his companion from Leoti was a serious field paleontologist, George recommended Dad for membership in the SVP. This organization of scientists, students, artists, and preparators promotes and regulates scientific activity related to fossils. Its purpose is to educate the public and to support the conservation of fossils and fossil sites worldwide. My parents were pleased to be able to travel to SVP meetings with George and Anna. The two couples journeyed by rail to New York; Chicago; Ann Arbor, Michigan; Lawrence, Kansas; and Denver, Colorado. During these trips, Anna and Mom explored the cities and browsed the museums while George and Dad attended scientific talks and met other paleontologists.

Dad understood the requirements for being a field paleontologist, but when he met museum paleontologists at the SVP meetings, he found dedicated scientists outside his normal frame of reference. He met people who valued scientific discovery and just happened to have colorful personalities. Some of them, such as Barnum Brown and Alfred S. Romer, were famous scientists and published authors. They were admirable, ardent men (and one woman). Membership in the SVP has since changed, and more women are now involved. In 2024, the past president and current president of the SVP were women: Jessica Theodor of the University of Calgary, and Margaret Lewis of the Richard Stockton College of New Jersey, respectively.[24]

At the New York meeting in 1956, William King Gregory (1896–1970) was a keynote speaker. Born in New York and educated at Columbia University, Gregory was a primatologist, paleontologist, and comparative anatomist. Dad heard Gregory's presentation in New York and frequently repeated his figurative illustration of the Empire State Building representing geological time. The metaphor went something like this: "Think of that building across the way, the

Empire State Building, as representing all the Earth's existence. Just under the roof would be the time when life on Earth begins. If you place a nickel on top of the roof, that is time since the age of the dinosaurs. Then if you place a cigarette paper on top of the nickel, that's the amount of time humans have existed." In the 1950s, scholars called the epoch that followed the last glacial period, when humans rose to prominence, the Holocene. A more recent designation for the cigarette paper epoch is the Anthropocene, a term coined by Dutch chemist Paul Crutzen in his short essay, "Geology of Mankind," published in *Nature* in 2002.[25] Some scientists use this term for the start of significant human impact on Earth (measured by biodiversity loss and climate change).

Dad met Alfred Sherwood Romer (1894–1973) and bought his book, *Vertebrate Paleontology*, for his library. He also met George Gaylord Simpson (1902–1984), who was a professor at Columbia University and curator of fossil mammals at the American Museum of Natural History at the time. Romer and Simpson were founding members of the society and its first two presidents. The highest award the SVP bestows is the Romer-Simpson Medal, for "sustained and outstanding scholarly excellence in the discipline of vertebrate paleontology."[26]

The successor to Barnum Brown at the American Museum of Natural History, Edwin H. Colbert (1905–2001) served as curator of vertebrate paleontology there for forty years. He was a leading authority on dinosaurs and authored twenty books. Dad met him at the New York SVP meeting as well and was influenced by his 1955 book, *Evolution of the Vertebrates: A History of the Backboned Animals Through Time*.

In an April 1961 letter on Fort Hays memo paper, at the age of seventy-eight, George wrote to Dad: "Got word yesterday that the powers that be will retire me July 1, 61. Old Age [his emphasis] the reason given. Let's go fossil hunting soon. G.F.S."

Before his retirement, George visited us often. Following his retirement, however, driving became challenging for George. One time, while driving his car into his garage, a freestanding wood-frame structure, he rammed it into the back wall of the garage. The problem was he stepped on the gas instead of the brake and shouted out, "Whoa!" He laughed about it afterwards, saying, "That car just didn't respond like the mules did!"

Orv says that one of his most entertaining duties when he was in Hays was driving George around. For the 1962 meeting of the SVP, held in Lawrence, Orv was the chauffeur driving "the dignitaries," George Sternberg and Myrl V. Walker, to the meeting. Orv remembers checking out a vehicle from the university and the president intoned, "*You* do all the driving. Don't let George behind the wheel!"

When Dad brought the plesiosaur to Hays, George was seventy-two, and he retired only six years later. He lived eight years after his retirement, the last two in a long-term care facility in Hays. He died October 23, 1969, at the age of eighty-six, and is buried alongside Anna in the Mount Allen Cemetery in Hays. FHSU's Museum, now located off-campus on Interstate 70, was named the Sternberg Museum of Natural History to honor George and the rest of the Sternberg family.[27]

Myrl Walker, Dad, and Orv served as pallbearers at George's funeral. It was a clear fall day when George was laid to rest. Standing in the sun at the burial, Dad gazed sadly at the headstone that bore George and Anna's names, thinking that now his mentor would become part of the ecosystem. Just then, a small whirlwind of leaves started eddying toward the grave, causing the entire funeral party to look up and take notice. Orv and Dad glanced at each other in wonder. The stirring wind settled down, then a single large oak leaf came to rest against the Sternberg tombstone.

George Sternberg received his original appointment to oversee the development of a museum on the Fort Hays campus in 1927. The original museum George opened was part of Picken Hall, the first building on campus. Then it moved to McCartney Hall, where it stayed for decades. The museum in McCartney was the one we Bonner kids explored while Dad was visiting with George in his lab. In the mid-1980s, the university acquired a large dome-shaped building that had been a health center on Interstate 70, on the northeast side of Hays. After seven years of planning and major donations from university benefactors, the Sternberg Museum of Natural History opened in 1994, two years after Dad's death.

Today's Sternberg Museum of Natural History, located on 3000 Sternberg Drive, sits on a rise visible from Interstate 70 West. The large, rectangular building with its round adjoining hall provides plenty of square footage for paleontology, geology, botany, paleobotany, and zoology displays and collections. The museum boasts life-sized restorations of extinct animals such as *Tyrannosaurus*, *Tylosaurus*, and *Pteranodon*. It even has a larger-than-life-sized restoration of George Sternberg himself, in khaki fossil hunting clothes and well-worn fedora, collecting a

Xiphactinus. George Sternberg's famous fish-within-a-fish fossil is still a main attraction, and so is the *Dolichorhynchops* Dad donated to the museum.

In his impressive tome, *Oceans of Kansas,* Michael Everhart, Sternberg Museum curator-at-large, clarifies all species that have been identified to date in the Cretaceous formations of Kansas and provides a historical overview of paleontology in the region. The book features six portraits of notable scientists who collected fossils in Kansas, beginning in 1866. The pantheon of paleontologists includes Edward Drinker Cope, Othniel Charles Marsh, Benjamin F. Mudge, Charles Hazelius Sternberg, Samuel L. Williston, and George Fryer Sternberg.[28]

But Everhart also gave credit to paleontologists who did not work for museums. He mentions others "who made significant contributions to our knowledge of the paleontology of western Kansas, but whose names are relatively unknown. ... people like Judge E. P. West, Joseph Savage, Handel T. Martin, William E. Webb, Marion Bonner, and others."

History Lessons
Late 1950s–1960s

FOR CHRISTMAS IN 1956, Mom gave Dad a blond Gretsch guitar. A self-taught troubadour, Dad had tinkered around on the guitar during the Whoopie Maker days and while in the Army. It was another way to perform. He could pluck out a few classical songs but favored railroad songs like "The Ballad of Casey Jones" and Jimmie Rogers numbers like "Waiting for a Train."[1]

The year of the guitar, 1956, was also the year of the piano. Pop purchased an upright piano from Anna and George Sternberg for twenty-five dollars that year. George had it delivered by pickup truck to our house in Leoti. Now our family could play duets with guitar and piano. Mom enjoyed listening to Chris, then me, practicing at the keyboard. She herself couldn't carry a tune and was amazed by Dad's ability—passed down to his children—to play songs by ear.

Dad hummed excerpts from Chris's piano numbers such as "Nola," by Felix Arndt, "Narcissus," by Ethelbert Nevin, Strauss waltzes, and Debussy etudes. To classical numbers he would add lyrics: "Come on and do the dishes," singing them loudly to tease Chris when she began practicing the piano right after dinner. (The crescendo at the end went, "Dishes are waiting for you to do! Dishes are waiting for you—boo hoo!") Mom admonished, "Oh, let her play. I enjoy listening to her!"

Half of us played some form of band instrument in senior high school: Letty the trombone, Steve the sousaphone, Chris the flute, and me the French horn. During his college days, Chuck started playing Dad's guitar and still owns it. Chris's piano lessons started with the arrival of the Sternberg instrument and lasted for four years. I grew up listening to both Chris and Dad pounding the piano and took lessons too, for seven years, starting in the second grade. And all of us Bonner kids took part in glee clubs and choral groups in school and could sing harmonies and pick out tunes on the piano. None of us took voice lessons, but no one needed them to burst into song. Dad, influenced by the movies, warbled while puttering or working on fossils, usually connected to his activity: "We dig,

dig, dig, dig, dig, dig, dig / In a mine the whole day through" or "I'm looking over a four-leaf clover / That I overlooked before."[2]

From his childhood on, Pop was a singer and entertainer. Adding instruments to the mix in this era opened up music performance to all of us. Also, before the 1960s, Dad really didn't have time to sit around and play guitar for hours. But after the post-war bustle settled down, after he sold the farm and the movie business went in decline, he did much more leisure strumming. That free time also led to more local history exploration and a surge in fossiling.

Whenever museums or individuals like Dad wanted to hunt fossils on privately owned land, they needed to first secure permission. Currently, museums make written arrangements with landowners, but in Dad's hunting days, he had hand-shake agreements with most of the ranchers of Logan County and Gove County.[3] They permitted him to come and go as he pleased, and he was very respectful, always treading carefully and securely closing gates. In that era, none of the pastures were padlocked. All the landowners knew what Dad's fossil-collecting vehicles looked like and would sometimes come by to see what he was working on. If they came into the pasture to check on their cattle, they said hello, but most often, once they saw it was him, they waved and drove on by, intent on finishing their own labors.

The group Dad called his post-war children enjoyed a wienie roast in the fossil beds in 1961. Left to right: Dana, Chuck, Chris, Melanie.

To visualize locations of outcroppings, Dad often referred to a line-drawn map issued by the Geological Survey of Kansas that indicated all the fossil beds in Gove and Logan counties. When I was in junior high, I used colored pencils to make the Pierre Shale dark green and the Niobrara Chalk light green. For the museums receiving specimens, he would send coordinates or land descriptions and indicate whether they were found in upper or lower Niobrara Chalk or Pierre Shale.

We Bonner children kept secret any knowledge of specific locations where fossils were found. Mostly, the locales were known by the family's place names, which were often colorful. If Dad said, "Let's go to Cedar Canyon," everyone knew where that was and recognized the bluffs when the car pulled up to the eroded beds. The Big Place, the Wishbone, the Valley of the Mosasaurs, Cedar Canyon, Hell's Half Acre, and Castle City were all Smoky Hill Chalk sites.

Pop bought plaster of Paris, a crucial ingredient in collecting fossils, from the local lumberyard. From the time he watched George Sternberg collect the *Xiphactinus* near Oakley in 1930, Dad tried to measure up to his mentor. Plaster tales (or perspectives) throughout the collecting years also involved teaching my brothers how to mix and pour it in the correct thickness—Orv and Chuck were his most interested students. I remember watching the process, fascinated. It was a favorite recipe. In a five-gallon paint bucket, Pop mixed water in the white powder, stirring constantly with his suntanned hands, the glossy, heavy batter turning into bright white gloves covering his forearms. When he was satisfied it was well mixed, he scraped the plaster off his hands and arms with a wooden paint stirrer, then splashed water over them to rinse off the rest.

That was what it looked like when it went smoothly. Orv remembers a plaster incident when the family returned to Leoti after World War II. On that outing, ten-year-old Orv found what was, up to that point in time, his best fossil. It was a partial skeleton of a *Pentanogmius*.[4] The family took a trip to the fossil beds to collect the five-foot-long rounded fish, and Dad was using the occasion to teach Orv how to mix plaster.

It was a chilly day when they set to work. They dug around the outline of the skeleton and situated the frame to prepare for the plaster pour. During the time that Dad had been away in the Army, the plaster had turned old, so when he went to mix it in his favorite bucket, it was gummy. He tried adding more water to get the right consistency, but it was simply bad plaster. According to Orv, our frustrated father "kicked the bucket down the hill."

The first fossil Chuck found that he says was "worthy of plaster" was in 1962 when he was twelve. Chuck's fossil was a *Cimolicthys* skull. He remembers watching the whole operation, Dad stirring the plaster in the paint bucket. When ready, he poured the smooth plaster carefully into the frame, taking pains to plane it into the corners evenly, using his hands and the stir stick to even out the surface. The square frame, when filled, resembled a large white sheet cake.

A *Cimolichthys* had an average length of three to five feet. Its jaws held three rows of sharp teeth. Resembling a modern barracuda or pike, this fish is actually related to salmon and has distinctive vertebrae with an hourglass shape in the middle and large, diamond-shaped scales on its back.

After taking the fish skull home, Dad kept it in his small lab at the back of the theater building. He allowed Chuck to carefully pick out the details of the fossil, exposing the bones of the skull until the specimen was clearly visible. In his own way, Chuck was hooked by a fish the same way Dad had been in 1925.

Later, there was a generational disagreement about the amount of plaster the collector should use. In Chuck's opinion, Dad often used too much, which made the resulting cast heavy and awkward. Dad, on the other hand, said Chuck "scrimped on plaster" and that made collected fossils too thin and fragile. The best use of plaster is in the eye of the collector, and the trick is to get it thick enough that it protects the specimen but not so heavy that it breaks the springs of the collection vehicle.

When I was digging through boxes of Dad's ephemera while researching this book, I found a tiny leather diary. The pocket-sized book was given out as a premium from Western Hdwe. & Supply Co. of Leoti. Dad wrote his name on the owner identification line and recorded events that occurred in 1961 and 1962. The first descriptions in it were about a ghost town called Sheridan:

> Located the town of Sheridan, Kansas, the end of the track of the Kansas Pacific R.R. (U.P.) in 1868, found bits of glass, cans, crockery, bits of iron, barrell [*sic*] staves etc. 35.3 miles N. [of Leoti] on 25. ... [Mom's handwriting] Jan 12, 61 located saloon—on E to bridge & end of line in '68. rifle pits on N hill, saloon is 9 x 12 paces S of tracks.

In 1961, Steve was a senior in high school, Chris was a freshman, Chuck and Dana were in sixth and second grades, and I was a preschooler. Since Dad closed the theatre in 1960, the family was able to attend evening functions, including Steve's basketball games. The entries for 1961 after Sheridan are about Steve's four-day basketball tournament held in Scott City, the town on Highway 96 twenty-five miles due east of Leoti. Then there is a long break in the diary before Dad starts recording details about fossils found in the summer of 1962, a productive time:

> June 19, 1962. Discovered complete specimen of *Gillicus*—57 vertebrae exposed, head separated from trunk. Chuckie discovered mandibles of *Empo*.
> June 20, 1962. Chuckie discovered complete *Ichthyodectes*? 64 vertebrae exposed, fins and ribs covered with chalk.
> June 21, 1962. Dana discovered skull & several vertebrae of *Empo*, took pictures of specimens. *Saurodon plebotomus*[5] [*sic*] complete fish, 92 vertebrae exposed, fins intact and present. Mosasaur specimen, *Clidastes*, caudal portion, 47 vertebrae, no paddles.

Empo is a name E. D. Cope gave in 1872 to a genus he thought was different from *Cimolichthys* due to the teeth. Since then, scientists have determined the earlier name *Cimolichthys* (Leidy, 1857) is accurate. Dad used the names he learned from Cope's book, and museums still carry the label "*Empo*" on certain specimens.[6] Dad called this type of fish Old Man *Empo* based on a character he knew from his childhood, Old Man Empey.

Dad was now querying Chicago's Robert H. Denison regularly. In one letter in 1961, Denison told Dad that the Chicago Natural History Museum[7] was interested in a *Pachyrhizodus* and "has approved payment of the shipping charges. If this is agreeable to you, we will make you an offer after we have seen the specimen." Denison accepted other fossils from Dad around the same time.

In addition to Denison, Dad made another connection at the 1960 Chicago SVP meeting—then-doctoral student Shelton P. Applegate. Applegate was learning about prehistoric fishes from the renowned expert Denison while employed at the Field Museum.

As for the *Clidastes* mentioned in the tiny pocket diary, we are not sure where Pop shipped this column of mosasaur vertebrae. It is possible he sent them to one of the museums that already held *Clidastes* material, to be used in a composite skeleton. In successive years, he would find complete skeletons of *Ty-losaurus* and *Platecarpus*, the other two prevalent genuses of mosasaur in the Kansas chalk, but a complete skeleton of

In 1960, Dad posed by the garage while crating a Pachyrhizodus to be shipped to the Field Museum in Chicago.

the smaller, swifter *Clidastes* mosasaur eluded him. *Clidastes* reached an average length of ten to twelve feet and probably inhabited shallower areas of the sea.

What got Mom and Dad so interested in finding Sheridan, the Logan County ghost town mentioned in the first pages of the tiny diary? That exploration of Kansas history started when Mom discovered a thick book published in 1872 called *Buffalo Land*. She checked it out from the Kansas Traveling Library, a bookmobile service that brought obscure works to rural areas. The Kansas Traveling Library Commission operated from 1899 to 1963.[8]

The densely printed volume with an embossed leather cover was a gold mine. An antique edition, it fascinated my parents because it detailed the end of the Old West in western Kansas, expounded on locations such as Forts Hays and Wallace, and even discussed familiar fossils. It was written by William E. Webb, the founder of Hays City (later called Hays), a man who had been a newspaper reporter during the Civil War, an adventurer, and a land manager for the Union Pacific Railroad as it was being built through western Kansas and eastern Colorado. Webb was responsible for selling the land given to the railroad by the U.S. government. He was elected as the first state representative for Ellis County in 1868.

Buffalo Land commented on buffalo hunts, the wild towns, and characters on the plains. A work of "boomer" literature, it encouraged readers to settle in the lands around the railroad's expansion. In 1869, the town of Sheridan was the terminus of the Union Pacific Railroad in northwest Logan County. The descriptions of Sheridan in Webb's book were so clear that my parents could locate the town, even though it was completely covered by buffalo grass.

Buffalo Land is a comic travelogue, complete with invented personalities, like a Wild West version of Dickens's *Pickwick Papers,* a much-copied literary form. In recent years, historians and scientists have taken note of a key character, Professor Paleozoic, and speculated the quasi-fictional person was Edward Drinker Cope.[9] Michael Everhart remarks that Webb and Cope were friends, and Cope even named a crocodile after Webb, calling it *Hyposaurus vebbii.* The two chapters in *Buffalo Land* on fossils were written by Cope and later published in scientific journals. Among his many pursuits, W. E. Webb acquired fossils, including the first known Polycotylid plesiosaur, *Polycotylus latipinnis,* in 1869. That specimen, described by Cope, ended up in two museums—the Smithsonian in Washington, D.C., and the American Museum of Natural History (AMNH) in New York.[10]

Pop himself made some guesses as to the identities of the fictionalized characters in *Buffalo Land.* He thought Professor Paleozoic was B. F. Mudge, that Doctor Pythagoras, M.D. was J. D. Parker of Topeka, entomologists Colon and Semi-Colon were University of Kansas professor Francis Huntington Snow and his son, and Tenacious Gripe was Topeka politician James H. Lane.[11]

Webb's *Buffalo Land* portrays the raucous town of Sheridan in vivid detail. Sheridan was the terminus of the Kansas Pacific railroad for nearly two years, and during that time, some two thousand people congregated there. *Ghost Towns of Kansas* describes Sheridan as a rough-and-tumble settlement that "was host to every bad element on the Kansas plains including gamblers, horse thieves, buffalo hunters, murderers, and prostitutes. Often it was not an exaggeration to say that Sheridan had a 'dead man for breakfast every morning.' "[12]

Webb, in an article in *Harper's New Monthly Magazine* in 1875, said of Sheridan, "We christened it after the gallant Phil, then stationed at Hays. When the general was introduced to his namesake, he remarked that, as a seat of war, it strongly resembled the Shenandoah Valley. ... 'I'll give you a high lot,' was a threat in Sheridan and meant six feet of soil on the hillside." Describing how the town grew up, Webb wrote:

> Sheridan was situated on the side of a desolate ravine. The ever-lasting plain embraced it. Two solitary *buttes,* named "Hurlbut" and "Lawrence," had been placed on guard over the region by nature, and looked as wretched and dismal as sentinels in a penal settlement. A month's hammering, and the new town was built. Before one street had been surveyed, however, the engineer was called upon to locate a graveyard. ... During the first week three of the inhabitants moved into that quarter, all going, as the phrase

has it in that country, "with their boots on." During the winter the number increased to twenty-six.[13]

When Dad and Mom found Sheridan, they located its Boot Hill as well. There they discovered human bones and reported these findings to the state agency in charge of historical sites, the Kansas Historical Association. The state was aware of the ghost town because when they excavated for a railroad bridge, skulls and other human bones "came rolling out of the hill," not from the Boot Hill cemetery but a hill beside the railroad bridge south of the townsite.

Using an early Polaroid instant camera, Mom took shots of Sheridan's hanging trestle, which Webb called "Jack Ketch," where frontier justice was administered in the absence of trees. Other photos show Dad's finger pointing out human bones, including a humerus, found near Sheridan's graveyard; and Twin Buttes (or Hurlbut and Lawrence), two small, pointed hills that overlooked the townsite. To me, the most interesting picture taken seemed to be of nothing, just a rise in the prairie with some snow on it. What the melting snow revealed, however, were curving, deeply embedded track marks of the Butterfield Overland Dispatch stage line heading toward Sheridan.

The book enabled Dad to connect with the eminent paleontologist Edwin H. Colbert, a prolific author who was then director of vertebrate paleontology at the American Museum of Natural History in New York. Dad thought Colbert would be interested in the book and loaned it to him. Colbert apparently kept the book for several months until Dad reminded him to send it back. Colbert's answer, dated December 8, 1964, read in part, "Under separate cover I am returning *Buffalo Land,* which I hope reaches you without delay and in good shape. I want to tell you how very much I appreciated your kindness in lending this to me, and you may be sure I enjoyed reading it ever so much. I must apologize for not having returned it before now. ... I would not dream of having you pay for the return of this book. ... The dollar you enclosed is being returned herewith."

Perhaps Dad gently prodded Colbert to return the book because he and Mom borrowed it from the Kansas Traveling Library. By the time he got it back, that entity was defunct, so Dad added it to his own library. I can picture my parents, the co-researchers, lying in bed at night, him reading to her. I can see them chuckling over Webb-written lines such as, "Hays City by lamplight was very lively and not very moral," or shuddering over the description of the ironically named "Lake Como" near Sheridan, "I bathed there once and touched my bare feet upon a cold face which slid aside in the mud."[14]

Dad's acquaintance with George Sternberg also gave new life to western Kansas archaeological discoveries he and Mom made during the Dirty Thirties. In the mid-1950s, George talked my parents into getting involved with the Kansas Anthropological Association. The KAA's first meeting was at Fort Hays State University in April 1955. For the next year's meeting, my parents selected their best specimens of Native American artifacts, and Mom created a display of them. The display was a three- by three-foot piece of plywood covered with blue felt, and Mom affixed arrowheads and other chipped-stone tools to spell out KAA and the year 1956.[15] The KAA's archivist, Mary Conrad, told me in an email in 2023, "Some of the early members also were interested in fossils, and George F. Sternberg and his wife were charter members. Sternberg also was the Second Vice President for the first two years."

Dad didn't join every year but only occasionally. The KAA's records confirm he was a dues-paying member in the first, fourth, tenth, and eleventh years of the KAA. Orville was also a member when he attended Fort Hays State University, and he gave a presentation titled "Vertebrates in Kansas Sites" at the 1964 KAA meeting.[16]

Mom was always an avid reader. She was greatly moved by Rachel Carson's exposé about pollution called *Silent Spring* when it came out in 1962. She also read a detailed yet whimsical biography of Barnum Brown written by his wife, Lillian, in 1950 and called *I Married a Dinosaur*.[17] It recounted the Browns' adventures collecting in the field in international locales. No doubt Mom identified to a small degree with the author, whose husband was a man thoroughly devoted to dinosaurs.

Over the years, Mom put up with unusual behavior from Dad. He was always a searcher but was also a risk taker who sometimes did rash things, like the time he got fed up with a hunting party and took a taxi home from Eads, Colorado, a distance of eighty-three miles. For the long cab ride, he hired a guitarist to play songs for the journey. (Orv confirms this story, and the musician probably taught Dad a few chords and numbers.) Pop risked the family's income with his gambling, but the habit sometimes paid dividends, such as when the winnings purchased materials for the large house in Leoti. He sometimes made poor decisions under the influence of alcohol. But she forgave his impulsiveness and loved him deeply.

By the early 1960s, though, she was delighted to be living with a man who, in this stage of their lives, had given up drinking and smoking. Dad was focused on remodeling the theater building so they could have its rental income. For a fun sideline, they had fossils and local historical research.

Ever since the Whoopie Maker and Prohibition days, Dad had enjoyed an occasional "snort"—whiskey and bourbon were his favorites. Alcohol was a part of his life and a lubricant of his storytelling. He was not a blackout drinker, but went "on a toot" periodically. However, for the six-year period from 1961 to 1967, the era I like to think of as his Golden Years with Mom, he did not touch the bottle. During those years, I was four through nine years old, and unlike the other kids, enjoyed a sanitized version of my father and saw firsthand his deep love for our mother.

As for Mom, she truly was one who abided all phases of life. I can't help but have a romanticized view of her. She was loving, temperate, and patient. When not cooking or cleaning, she kept her hands busy with knitting and crocheting. One time when I was in second grade, I forgot to tell her about a neighbor friend's birthday party and called her from school in a panic. She swept to the rescue, knitting a Barbie doll outfit in a couple of hours, wrapping it, and bringing it to my friend's party after school.

You can't hunt fossils without water. Dad "portaged" water in old paint cans and gallon milk jugs when he wanted to mix plaster. But humans must have it to survive in the environment. For family outings, Mom created her own version of an insulated bottle, about which Dad crowed, "Other patents pending." She sewed fitted, thick green canvas coverings with zippers for quart and half-gallon water jars. The night before a hunt, she filled the jars, moistened the outer canvas, zipped them up over the jars, and chilled the bottles in the refrigerator overnight. They kept their cool (somewhat) in the field if they were placed in the shade. On scorching summer days, the biggest concern was that the contents of the jars were wet, so being slightly cool was a bonus.

This was a perfect example of Mom "making a silk purse out of a sow's ear," a habit learned from times of deprivation during the Great Depression and the Dust Bowl. She imparted that habit to her children. My very first memory of life in Leoti is of walking up and down the alley behind our house with Dana, searching for bits of colored glass that Mom could use in her melted-glass creations, which became earrings, cufflinks, and pendants. She owned a table-top kiln for this artistic repurposing.

After Dad parked the car at the top of a pasture overlooking the chalk outcrops, we kept the windows open and placed Mom's covered water jars in a shady area under the vehicle. Then when lunchtime came, we passed the jar around. While out away from the car hunting fossils, I remember sometimes placing a smooth quartz pebble in my mouth to keep thirst at bay as I roamed the beds. Upon returning to the car, we fossil hunters would drink up. While everyone slaked their thirst in turn, Dad sang, "All day I've faced a barren waste / Without the taste of water, / Water, water, cool, clear water."[18]

In the 1960s, we acquired an insulated metal chest, an old-style cooler. It was green and bore the words "Pleasure Chest" on its side. Long before the days of Yeti coolers, this was an excellent way to keep food and drink cold. But it was heavy and awkward, designed for the use of large groups. When Dad went out alone or with one or two others, he kept sandwiches and other food in plastic bags. One technique he employed for group lunches was to place peanut-butter-and-jelly or bologna-and-cheese sandwiches in their original bread sack (Dad's favorite was Roman Meal) until it was time to eat. Everyone was eager to see that Roman Meal bread sack come out at lunchtime.

Any food consumed in the fossil beds is tasty; exertion in this environment creates a deep hunger for refreshment and nourishment. The addition of salt alone—to hard-boiled eggs or carrot and celery sticks—can make your body go from feeling weak to satisfied. On one of her summer visits, Clare Jane introduced us to Little Debbie oatmeal cream pies, which were truly gratifying.

Mom instructed us to forage for materials she then turned into art projects. But sometimes she asked us to scavenge for edible plants. Mom made jelly from the wild currents and choke cherries that grew in the ravines between chalky outcroppings. One of Mom's favorite alliterative questions was, "Did you check the choke cherries, Chuck?" Sometimes the jam in the sandwiches we ate in the fossil beds was made from the fruit of the bushes growing nearby.

Chris, one of the best foragers, was influenced by our parents' appreciation of the arts: Mom's yen for painting and crafting, and Dad's talents in music. In high school, she gravitated toward music and drama. No art courses were offered during her era, but she was called upon to draw posters and work on stage backdrops and other projects. A 1964 graduate of Wichita County High School, in addition to band and vocal performance, she took part in drama, playing the lead in *The Diary of Anne Frank* in her senior year.

One of my most haunting childhood memories occurred while attending that play in the fall of 1963. A fidgety six-year-old, I was having a hard time sitting still in my folding chair and paying attention. The red-curtained stage was at the end of the darkened gymnasium. All of a sudden, I heard my sister's scream piercing the darkness. I sat up, terribly upset, then became mesmerized as the curtains opened and I realized her agony was a role she was portraying. I stayed more aware of the play after that, particularly the parts when Chris was performing the role of Anne Frank.

Chris eventually earned her fine arts degree from Fort Hays State University. As an adult, she looks back on our upbringing with veneration. "I am so glad we had the parents we had," she affirms. "They were both so unique and talented. I was in awe at how much they both knew and taught us all."

I have warm memories of Chris taking care of me when I was little. She would perch me on the back of her bike and pedal to the drugstore on Main Street, where we could get single-scoop ice cream cones for a nickel. I always asked for lime sherbet. It seemed like such a luxury. Where she appreciates how much Mom and Dad taught us, I also appreciate everything my older brothers and sisters taught me, a storehouse of knowledge and experience.

California Connection

1960s

CLOSING THE PLAZA THEATER brought two bonuses to my siblings. First, they were no longer required to run the show or clean up. Second, they could devote more time to their extracurricular activities, such as music, drama, and sports, and Mom and Dad could finally watch them. Up until his senior year, my brother Steve had been running the movies except for nights when he had games, and then Dad stepped in (and wasn't happy about it because the ticket sales barely covered the film rentals). Now Steve was gratified that our parents and his younger siblings could see his performances as starting forward for the high school basketball team. High school games were the highlight of the week in small towns like Leoti.

After Steve graduated from high school at seventeen, he enrolled at Fort Hays State University. Unfortunately, he didn't pay much attention to his classes. Instead, he partied and played cards with his friends, habits he learned from Dad. He also frequented the Hays pool halls at night, gambling on games of snooker. He disenrolled from FHSU and decided to try Kansas State University in Manhattan next. At that time, Clare Jane, her husband, Ray, and two children, were living in married student housing near the KSU campus while Ray finished his doctorate in veterinary medicine.[1]

Steve visited the Askeys often during that semester at K-State but still wasn't sure what career to follow, so he put college to the side and enlisted in the U.S. Navy. Steve enjoyed fossil hunting but didn't consider paleontology a potential career, believing that was an area Orv had claimed. In some ways, Dad had pushed Orv into paleontology, much as some parents push their children into law or medicine.

In the summer of 1963, the whole family was hunting the Big Place—a large group of badlands in western Logan County. It was a normal day for everyone—all the youngsters scattered out among the ravines, hoping to be the first to find a shark's tooth. In just a few moments, we heard Orv yell triumphantly, "Shark's *toooooth,*" winning the competition and dashing our hopes for bragging rights.

Steve, on leave from the Navy after boot camp, felt himself getting nostalgic as he walked up a gulley. He belted out a song from one of his high school musicals, "I am the captain of the Pinafore, and a right good captain too," while scanning the cliffs in front of him. Halfway into the song, he noticed some strange vertebrae leading back into a bluff. As he had learned to do, he started unearthing around the fossil, then slowly and methodically scraped with a Marsh pick from above the fossil down to the layer where the specimen was going back into the chalk wall. He dug enough overburden off the fossil to see that the vertebrae were flat and overlapping like poker chips. Feeling a burst of enthusiasm, he looked west to where Dad and Orv were digging on another type of fish. "Dad! I think I've found a shark!" he yelled.

Orv responded first, greeting Steve's announcement with skepticism, but started moseying up the hill where Steve was working. It didn't take him long to concur that the find was unusual, and that the shark could be there, heading back into the rock. Steve had found the shark by the caudal vertebrae and had started digging toward the rest of its body. A complete specimen would be an extraordinary find because sharks, being cartilaginous, generally did not fossilize well.

Hearing their shouts, Dad went over to have a look. He was excited by the prospect and took over the digging. When he uncovered a succession of straight silver-dollar-like vertebrae going into the wall leading toward the middle of the body and the head, Dad got more excited. The next big concern was: Would this be vertebrae only, or would the digging culminate in a complete specimen? Dad marked the site and said he would come back after lunch to finish excavating it. When he resumed digging, he was delighted to see that the disarticulated vertebrae in the large middle portion of the shark ultimately led to a skull filled with serrated teeth.

Dad contacted a shark expert at the Los Angeles County Museum of Natural History, Dr. Shelton P. Applegate (1928–2005), or "Shelly." Shelly had been a graduate student at the Chicago Museum of Natural History when Dad was sending fossils there. Applegate was extremely interested in the rare, nearly complete shark. After Pop collected it, using the frame-and-pour method, he shipped it on spec to Los Angeles in a wooden crate made large enough to cradle and protect the fossil. The large boxes made nice nests for the fossils, packed in cloth and sometimes parts of rubber tires for added protection. He carefully wrote on the plywood the museum's address and "Fragile—This Side Up" with large permanent markers. This was before UPS and FedEx were firmly established shippers, so the fossil, like those of Cope, Marsh, and Sternberg, traveled to its museum destination by rail. Shipping by freight train directly from Leoti to Los

Angeles was not possible at the time, so Dad took the fossil to Oakley, seventy miles northeast of Leoti. (Shipment from Oakley also stopped being an option in the late 1960s.)

The *Squalicorax* was the most complete ever found. It was the first of many specimens Dad sent to Shelly starting in 1963. When Shelly and Dad spoke on the phone, Shelly said in his Virginia accent, "I'll take any old sharks [pronounced "shocks"] you have!" to which Dad replied, "Would you like those shocks to be 220 or 440?" Steve's *Squalicorax* opened the door to other quality Kansas fossils that would go to Los Angeles in the coming years.

Steve's Squalicorax is articulated at the tail and skull ends and scattered in the middle. Courtesy of the Natural History Museum of Los Angeles County.

Shelly Applegate and his family visited Leoti and the chalk beds twice—once during the summer of 1962 as Shelly was moving from Chicago to his new position in Los Angeles, and again in 1965, when he hoped to find more sharks. Unfortunately, Steve's shark was a rare find, so they mainly located partial skeletons of teleosts, or bony fishes, and some *Pteranodon* material. Mom and us rangy Bonner children enjoyed the visits from the Applegate family and showed the "city folks" the ropes around the fossil beds.

Steve's Navy service was principally on aircraft carriers. He served from May of 1963 through May of 1967. He remembered how one time in 1966 the crew was boarding the USS *Bennington,* and he looked down into the water to see, sculling just under the surface, an enormous Great White shark. Even though it was a ninety-degree day in the South Pacific, Steve felt a chill as he looked at the massive predator, which appeared to be about twenty feet long. Some of his fellow sailors were crying out, "Shark!" and pointing down to the water. Steve smiled to himself, remembering how Dad had quipped, "When a sailor hollers 'Shark!' you'd better pay attention!"

After finding the *Squalicorax,* Steve's life took an entirely different direction. That shark was the highlight of his fossil discoveries. In 1966, toward the end of his service in the Navy, he married Sonya Bristow, the sister of Letty's husband,

Bud Bristow. After his discharge from the Navy, Steve took a summer job in Oklahoma while Sonya finished her bachelor's degree in 1967; then the couple headed to Lawrence. While there, Sonya attained her master's degree in social work and Steve his chemistry degree from KU.

Steve had served in the Navy from the year after the Cuban Missile Crisis to the start of the Vietnam War. He was on the University of Kansas campus when students all over the country were protesting the war. At KU, the most extreme of these protests involved the bombing of the computer center and the burning of the Student Union in 1970.[2] Steve didn't take part in the militant activities that threatened to shut down the university, but he got some free T-shirts that were water damaged after the Student Union fire.

The early 1970s saw an expansion of several scientific departments at KU, and it was there that Steve spent most of his time. His work at CRES (the Center for Remote Systems)[3] in the complex west of the main campus introduced him to computer processing of scientific data. This was good preparation for the job he was offered in the oil industry right after graduating. He understood the geology behind the oil business but was lured into his lifelong career with Mobil Oil by a position in electronics. At Mobil, he worked with supervisory control and data acquisition (SCADA), a system architecture linking computers with networked data communications for high-level supervision of machines and processes. In the mid-to-late 1970s, SCADA in the oil industry led to safer underground pipelines with better controls. Dad didn't really understand what Steve's job entailed, but told everyone proudly that Steve was "in computers."

Dad's friendship with Shelly bloomed during a time when the Los Angeles museum was experiencing a period of growth. Shelly was another spirited figure in the world of paleontology. Applegate served in the Navy (where he saw plenty of modern sharks) and received his bachelor of science degree from the University of Richmond, his master's from the University of Virginia, and his doctorate from the University of Chicago. At the time of his death, he was a professor at the University of Mexico in Mexico City and an associate professor at Harvard University in Boston, where he was "a well-known expert on living and fossil sharks." In addition to Mexico City and Los Angeles, Applegate worked in museums at the Smithsonian, Duke University, the University of Chicago, and Arkansas State University.[4]

For some reason, visiting paleontologists tended to be outsized characters. Shelly was no exception. He was a renegade in academia, pursuing new challenges and moving frequently to different museums. In each curatorship, he made a profound impact on the graduate students who worked with him. In *Curators*, Lance Grande of the Field Museum paints a vivid portrait of Shelly's work in Mexico with new species of fossil fishes in the state of Puebla. He knew Shelly toward the end of his career, whereas Dad met him as a younger man in the 1960s. Grande said of him:

> Shelly was one of the most unusual scientists I have ever known ... never afraid to go out on a limb with his ideas and opinions. Large, loud, and white haired, he resembled a beardless Santa who was still wearing clothes that he had slept in the night before. He was a free spirit with a lust for life and a pragmatic attitude about how to live it. He was short on inhibition and always up for a good party.[5]

It's no wonder Shelly and Dad hit it off. It also helped that Shelly was eager to acquire Dad's fossils. At seventeen years Dad's junior, Shelly didn't fill a mentor status the way George Sternberg had, but he was an academically trained advisor who gave Dad the latest information about all kinds of fishes. Fittingly, in 2014, scholars posthumously named a megamouth shark after Applegate. DePaul University paleoichthyologist Kenshu Shimada, who earned his master's at Fort Hays State University, co-authored a paper naming a 23-million-year-old shark found in California's San Joaquin Valley *Megachasma applegatei* in honor of Applegate's lifetime devotion to the study of sharks.[6]

The Applegates weren't the only museum family to visit us in Leoti. Another paleontologist with wife, youngsters, and trailer in tow was Gordon Edmund of the Royal Ontario Museum. The Edmund family came through western Kansas on their 1961 summer vacation. The Edmunds spent a few days running around in the chalk with us Bonners but didn't find anything significant on that trip. Instead, we played games in the chalk, and the Canadian kids endured the heat with fortitude. Dad listened intently to Gordon's explanations about his specialty: fossil reptiles.

During their visit, Dad teased the Edmunds playfully about their Canadian accents. While playing in Chuck and Dana's room, the Edmund boy thought his sister was bothering them too much, and he shouted at her, "Get oot of this hoose!" That expression immediately caught fire and was repeated frequently.

Dr. A. Gordon Edmund served as curator of vertebrate paleontology at the Royal Ontario Museum from 1954 to 1990. Known affectionately as Gord, he earned his doctorate in vertebrate paleontology from Harvard in 1957. In 1965, Edmund was included in an article on the museum that was published in *MacLean's* magazine. In the story, the writer described behind-the-scenes activities: "Like an iceberg, much of any museum is hidden from public view." Collections under study included vertebrate paleontology material. "Dr. A. Gordon Edmund, in charge of the dinosaur galleries, is currently studying thirty-five reptiles (the nearest living descendant of the dinosaur) in a ROM laboratory, charting the rate and rhythm of their tooth replacement. Since one reptile may replace each tooth five times a year and has twenty-five teeth to each of its four jaws, Edmund is studying five hundred teeth in each specimen."[7]

One warm morning in 1965, when all of us kids were in school, Dad leaned over to Mom and said, "Let's take a peer at the Pierre today, Honey!" He had in mind a trip to an area west of Hell's Half Acre.

In Kansas the Pierre Shale (pronounced "peer," like the capital of South Dakota)[8] overlays the Smoky Hill Chalk, which means of the two late Cretaceous sediment layers, the chalk is oldest. Although many Bonner fossils came from the chalk, important specimens were also found in the shale. For those accustomed to hunting the yellow chalk, where bones of dark and light blue-gray stand out, the eyes must adjust to hunting the Pierre. Instead of looking for darker bones in the matrix, you adjust to lighter ones against the gray, which ranges from a dark, gunmetal color to a lighter weathered gray. Spotting gray fossils on gray sediments is tricky.

After they got to the site, Mom decided to go off the beaten path. She ambled slowly to some unimpressive, small washes on the west side of the outcrops. Scouring the shale, her eyes landed on several shark's teeth. "Marion, come look at all these shark's teeth!" she called across the pasture floor. "They're white! And they have little side points!" He strode over to where she was scanning the rocks along the gully for teeth and bones. "Honey, you've found more than teeth here," he crowed. "This is a *Lamna* shark, Margaret. You've found something *rare!* Let's see how much of this shark is here." Mom stood up and kicked a long leg toward heaven. "Whoopeee!" she hollered. After Dad unearthed the specimen and saw how much was there, he too was excited. It appeared to have an almost complete skull. He teased her later about "kicking up," Rockettes style, over the fossil.

Years after Dad sent it to L.A., scientists determined that it was the most complete specimen of *Archaeolamna* to date. Up until this discovery, this genus of shark, also called mackerel sharks, had been identified by their teeth alone since bone matter was so scarce. Mom's shark had layers and layers of teeth, jaw and skull material, and a few weird vertebrae. It is now an important museum study specimen and over the years has been analyzed by shark experts. This was definitely a fantastic "shock" to send to Shelly in California. But Shelly did not realize its full importance at the time. The genus *Archaeolamna* was not named until 1982, and this specimen was not described until 2011.[9]

It wasn't the first time that Mom had channeled her inner Rockette. Earlier that month, she had located some neck bones of *Hesperornis*, the flightless swimming bird of the Cretaceous. She kicked up over that find too. Maybe her eyes noticed the three cervical vertebrae in the gray shale because they reminded her of the bones of turkey carcasses she had handled in her years of preparing holiday dinners.

Dad was starting to have a warm feeling about the Pierre. Granted, it was difficult to hunt in. The gray shale was shot through with gypsum, which glinted in the sun. But the more he and his family hunted the Pierre Shale, the more he realized it was a good source of unusual fossils.

Paleontological and historical outings were always a treat to Mom, giving her a break from her extremely busy life. She had a large family to feed, always aided elderly neighbors, and immersed herself in her church and clubs. She was also continually engaged in creative activities: painting, sewing, tin-can art, glass crafting, flower arranging, quilting. When it came to fossils, she had already shown a talent for preparation and would have done more of that if she had had the time and if Dad had encouraged it. However, she probably wasn't interested in more than dabbling, seeing paleontology as Dad's domain. On these occasions, with her children out of her hair, she demonstrated an uncanny ability to find rare fossils.

On October 10, 1963, after Shelly Applegate moved from Chicago to Los Angeles, he wrote Dad a letter that read:

> I hope you have had a productive summer. As to the cretaceous fish in our collections we really need everything but mostly specimens for exhibition. If you could send me a list of what you have on hand or the actual specimens, I am sure we can find the means to purchase them; however, we must have the specimens here. We are planning a Mesozoic Hall and I would like to have a large *Portheus*

[*Xiphactinus*] and several other complete fish. Any birds or reptiles other than *Mosasaurus* could be utilized. I hope to hear from you.

That encouragement was music to Dad's ears. Instead of a list, he shipped fossils. Between late 1963 and early 1965, Dad sent Shelly all fossils that he deemed museum-quality. After receiving them, Shelly prepared his own detailed list to give to the L.A. museum registrar for payment—thirty specimens in all. Most of the fossils were fish, but bird and *Pteranodon* fossils were also included. Starting with Steve's *Squalicorax*, the list (quaintly typed on a manual typewriter that needed to have its keys cleaned) also included a nearly complete *Xiphactinus* and a stellar *Pteranodon* skull, both found by Chuck. It is clear from his explanations of the specimens that Applegate was highlighting their importance to convince the museum to purchase the fossils.

On the first page is the museum's conclusive statement: "Approved by the Board of Governors, February 2, 1965. The Marion C. Bonner collection of fossil reptiles, fish, and birds: Thirty beautifully preserved specimens ... from the Cretaceous chalk, including a very rare shark, two types of flying reptiles, and parts of primitive birds and excellent fish. See attached list."

That list, which the Los Angeles Natural History Museum's current collections manager of vertebrate paleontology, Dr. Samuel McLeod, shared with me while I was researching this book, included details of the thirty specimens (*Squalicorax, Anogmius, Stratodus, Pachyrhizodus, Syllaemus, Albula, Gillicus, Cimolichys, Ichthyodectes, Pteranodon, Nyctosaurus, Protosphyraena, Saurodon, Portheus, Ichthyornis,* and *Hesperornis*). Standouts included Steve's *Squalicorax*: "An extraordinary fossil shark, *Squalicorax*, with vertebral column; skull with teeth in place; length of specimen is about ten feet. This shark is the Cretaceous ancestor of the Basking Shark. There is no similar shark as well preserved as this specimen in any museum in the world."

Seven separate *Pteranadon* specimens, collected from different sites, were used in the composite *Pteranodon* that is on display in the museum today. Applegate defined all seven *Pteranodon* specimens separately (noting the size derived from wing material was about seven feet long with a wingspread of about thirteen feet). He said of the large skull: "One *Pteranodon* skull complete, in an excellent state of preservation, lacking about three inches of the bill points. One or two cervicals [are] present, very unusual specimen, and well-articulated. I believe it is one of the best skulls that has ever been collected. Length four feet."

Shelly and Dad had joined forces to fill the California museum with Kansas fossils. Evidence of the Applegate family's visit to western Kansas in 1965 is found in the list's explanations. Applegate confirmed that one of the fossils, an excellent *Nyctosaurus* (a small pterosaur) was found by his wife, Anne, during the Kansas visit. Evidence of Shelly's amity toward Dad is seen in the fact that Anne Applegate found the fossil, but Shelly wanted to pay Dad for collecting it and included it on the list of Bonner fossils. The "Bonner Collection" name for the wall of fossils in the L.A. museum likely originated from Shelly's list. That exhibit, on display from 1976 through 1987, eventually made way for the museum's permanent Mesozoic displays in Dinosaur Hall, the current home of many Bonner fossils.

Chuck's nearly complete Xiphactinus, shown from above during Orv and Dad's excavation, is now at the Natural History Museum of Los Angeles County. Photo by Chuck Bonner.

One weekend in 1966, when Orv and his wife, Ellen Jean (Wilson) Bonner, were visiting the family, everyone packed up the car and went to check out the beds south of Lone Butte. That Saturday, Chuck located a large *Xiphactinus*. He spotted its tail fin, and as Dad and Orville dug back into the butte, they were surprised and pleased when the fossil turned back on itself. It had died and fossilized into a *U* shape. Chuck was excited. He was sixteen years old, and when the family returned to the site on Sunday, he brought his Brownie Hawkeye camera and snapped several photos of the excavation.

This *U*-shaped *Xiphactinus* was the *Portheus* Shelly Applegate included in the list of fossils the museum purchased. It is the only *Xiphactinus* found by the Bonners in the collections of the Natural History Museum of Los Angeles. In the list of specimens, Applegate defined this immense fish as a "Complete specimen with the exception of one-half of one tail lobe. Trunk articulated and curved; skull in palatal view; a little disarticulated but complete. Pectoral fins complete, disarticulated; skull drifted an inch or two from trunk and taken in burlap and plaster to enable its being rearticulated with trunk; well preserved specimen about fourteen feet in length. Will make an impressive and beautiful mount."[10]

By early 1967, Pop had accumulated another batch of fossils to send to Shelly. He kept most of them in the garage at Leoti but placed some of the more delicate ones on the library table near the kitchen. The years were productive for fossil hunting, and the rent money on the remodeled theater building gave Dad a modest, steady income. In addition, Mom put her talents to work by taking in sewing for a fee. (The only time I ever heard my mother curse was when she said "Damn!" after breaking a sewing machine needle.)

It was also in this era that Clare Jane and her husband, Ray, gifted us with a significant addition to the family. The Askeys had a black miniature poodle named Bridgette Antoinette that they had bred. From her litter of puppies, they chose one for us and sent her on an airplane from South Dakota to Garden City. Mom and Dad went to Garden City to pick her up. She was an instant hit, a little black cuddly mop. We christened her Dolli Antoinette.

We had owned outdoor dogs and quasi-feral cats for years, but now we finally had an indoor dog. Dana and Dad immediately set to house-training her. She became a member of the family and adapted easily into a fossil hunting hound. Dolli grew to a height of about fifteen inches at the shoulder and could hold her own with all of us. We rarely gave her a poodle cut—most of the time her fur was a heavy thatch that looked more like a puli than a poodle. The sweet little canine became an emotional support animal for us in the years to come. She was our first indoor dog and is probably the reason why I have always had them as household companions; since leaving Leoti, I've owned twelve dogs, mostly rescues.

Dad's biggest shipments of fossils to Los Angeles were in 1965 and 1967, but he peppered Shelly with fossils for years and spoke regularly with him on the phone until Applegate left the museum for a position at the University of Mexico in 1975. The major haul of Bonner fossils held by the Los Angeles institution, the largest natural history museum in the western United States, would provide data for years of future scientific research. Like the "Bonner Wall" at the museum, Dad wouldn't be able to see this research himself. Paleontologists studied the specimens and published their discoveries long after Dad's bones were interred in the Holocene (some might say Anthropocene) Epoch.

Storms and Sadness
Late 1960s

POP DROVE MANY USED cars over the years. Documented in the family photo albums from the 1920s to the 1960s are a Model T Ford, a black Studebaker, a 1936 Ford "fan," a white Chevy "ghost," a large blue Oldsmobile, two green-and-white Olds Ninety-Eights, and a sleek coppery tan 1959 Chrysler De Soto. Whenever we got in any car, he would say, "I'll drive," and we always shook our heads, because no one ever drove but him. Mom drove around Wichita County to run errands, but Dad handled longer trips.

In 1963, he bought a 1949 Chevy Suburban, a "perfect fossil wagon," from a local plumber for eighty dollars. He dubbed the vehicle "Spiker" after its previous owner. The back seats were stripped out, so we kids either sat on pillows on the floor or on a long library bench inside the cargo area. Dad put Dana and Chuck to work scraping the old paint off with razor blades, then painted it solid gray with leftover oil paint, a great color, he said, because it would blend with the fossil-bed scenery.

Chris graduated and married in 1964 and moved to California. Her departure from the nest left Dad and Mom to contend with only three of us. Chuck was fourteen, Dana was eleven, and I was almost seven. Because Chuck and Dana shared a room and I moved into Chris's room, Mom and Dad decided to close the door to the old part of the house, the Other End, to save on heating and electricity. My room shared a wall with my parents' room, and the boys were across the hall from them.

Chuck and Dana were always competitors. They threw walnuts at each other in the back yard and laid out a boxing ring in the side yard. Obsessed with trying to beef up their skinny bodies, they made weights out of coffee cans, plaster, and plumbing pipes for their "own home gym," a phrase Dad teased them about in a nasal voice. But he supported their efforts, installing a punching bag for them in the Other End. Pop demonstrated how to rattle the bag with a couple of different rhythms, one of them involving elbows. Every time he did it, I would think, "Anybody here wanna *box*?" The piano was also in the Other End, so Dad would sometimes light the gas heater in winter months to encourage my practicing. The Sternberg piano was central to my life in Leoti. While tinkering around with sheet music from multiple eras, I spent a lot of time in my own head, in piano daydreams.

On one trip to the fossil fields in 1966, Mom stayed home for one of her many club activities, no doubt happy to get a break from us. Dad decided to take us to the Wishbone, in Logan County. Going north out of Leoti on Highway 25, the smell of freshly turned soil turned to the smell of the successful local feedlot, and then as we passed into Logan County toward the Smoky Hill outcrops, the smell turned drier, grassier, chalkier. I still have this sense memory.

He divvied up the fossil tools and brushes from the back of Spiker, and the Coyote and the Mountain Goat lit out as they always did, searching for big fossil game. I stayed close to my father, listening to his songs and stories as we looked around, hopeful. We never went into the fossil beds and didn't feel hopeful.

It wasn't long before the Coyote hollered, "Dad! I think I've got something good over here! It's big and weird." Pop gave me an amused look, blue eyes widened in mock shock, and we took off to see what Chuck had found. Dad dug on it for most of the morning and was thrilled that it was a *Pteranodon* skull, the largest he had ever seen. He decided to collect it that day, so after lunch, he scoured the area and dug back deeper into the bluff to see if there was any more material. There were some cervical vertebrae with the skull, but no other bones could be located.

This was the large *Pteranodon* skull Shelly Applegate included in his 1965 fossil list. It was an important part of the wall of Cretaceous fossils displayed in the Los Angeles County Museum for many years, and it's on exhibit today alongside a composite *Pteranodon*.

That afternoon, Dad finished prepping the skull for collection, carefully applying a burlap-and-plaster jacket around it. Chuck and Dana helped, running to Spiker to get extra gunny sacks to encase the fossil. As the plaster hardened, thunderheads began rising on the eastern horizon. White cumulonimbus tower clouds could sometimes peter out in the heat radiating off the plains, but these were dark blue and ominous. The smell of the air changed from that of chalk to pre-thunderstorm ozone. "Looks like we've got a storm brewing," Dad said. "Let's get this baby out of here."

He pried the fossil from the rock matrix, and the boys helped him carry it up to Spiker. By that time, the atmosphere was turning dark, and large drops of rain started pelting us as we gathered up our tools and got into the Chevy van. Chuck and Dana sat in the back with the skull, and I hopped up front in the passenger seat. Dad ground the long gearshift into first, and we got out of the pasture as quickly as we could, just as the deluge hit. As lightning flashed around us and thunder boomed, Dad smiled at me and said, "Sounds like the tater wagons are rolling."

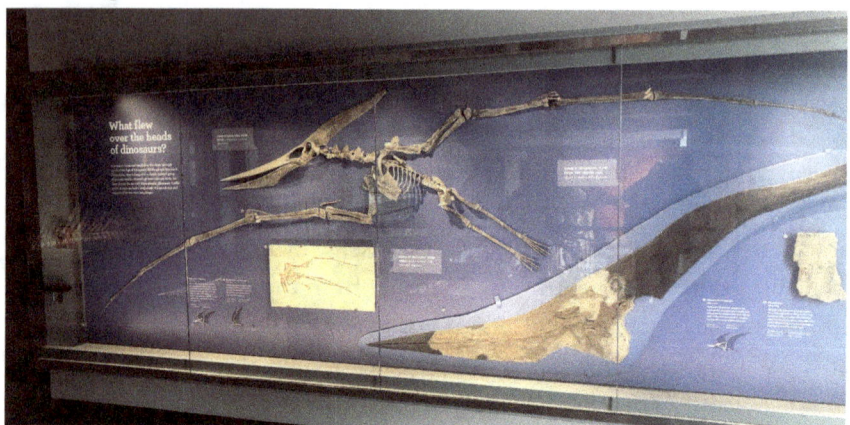

Chuck's large Pteranodon skull, right, in Los Angeles display; the composite Pteranodon was made from mainly Bonner fossil material. Photo by J. D. Stewart.

It was the scariest thunderstorm I had ever experienced. Dad said, "If this gets too bad, I may need to pull over until it passes." We got a mile or two down the sanded road and couldn't see anything out of Spiker's small side windows. It was like being in a deep blue car wash with lightning cracking in regular intervals. "Let's stop here for a bit," Dad said, pulling to the side of the road. He didn't want to turn off Spiker's engine because he was afraid it wouldn't start again, and we would be stranded. Spiker idled roughly, then it started "missing," one cylinder at a time. We sat there waiting, the wind and rain raging against the heavy frame of the fossil wagon. After a time, the engine got down to one piston firing: *clunk, clunk, clunk.*

As the band of thunderstorms moved over us and deep blue sky peeped out from the east, Dad gave Spiker some gas and adjusted the manual choke and throttle knobs, which were labeled *C* and *T.* The other pistons started hitting again, very slowly. The roads were drivable, so we made our way back to Leoti, discussing how we had lucked out: We didn't get stranded, and the rare fossil was safe. It had been daytime when we left the pasture but was evening when we got back on Highway 25. We ran into the house, all abuzz, telling Mom everything that had happened. Her large brown eyes looked worried. After we described the adventure, she told us, "You kids should *write* about your experience with the *Pteranodon* and the storm." I don't know if Chuck and Dana wrote anything down (if they had, they would have made graphic novels), but I ran to my room and wrote an account of the storm. Mom and Dad were pleased with it. That tale is gone, but now I have another chance to tell the story of the stormy skull.

Flying reptiles were a fantastic find. Toothed birds were even more so. The two main genuses of early birds found in the Smoky Hill Chalk were *Ichthyornis* and *Hesperornis*. Dad always yearned for bird material, and he turned the scientific name *Hesperornis regalis* into the hopeful phrase, "We will have *Hesperornis, regardless.*"

And so we did. In 1958 he found the most complete *Hesperornis* to that date and took it to the Fort Hays State Museum. That fossil is complete except for the skull. *Hesperornis* ("Western bird") was a large, flightless bird similar to a loon or cormorant. It had huge feet adapted for swimming and small teeth designed for catching fish. It reached a length of five feet and probably only left the water to lay eggs. The type specimen of *Hesperornis regalis* was named by O. C. Marsh.[1] (Pop's *Hesperornis* is shown on page 101.)

Currently, two of the most respected experts on *Hesperornis* and other extinct birds are Dr. Alyssa Bell and Dr. Luis Chiappe of the Natural History Museum of Los Angeles County. They included another of Dad's specimens from the Sternberg Museum collections in their research on the diversity and distribution of Hesperornithiformes, an order that encompasses *Hesperornis* and *Baptornis*. In addition, *Hesperornis* material that Chuck found in 1982 and that went to the KU museum—a nearly complete skull and a few limb bones—is included in their comprehensive analysis.[2] Chuck's *Hesperornis* discovery was exceptional because it was the first skull to show a recognized predentary bone, which allows the lower jaws to move apart in the process of swallowing prey. This feature of *Hesperornis* was the subject of a paper by Dr. Larry Martin, curator of vertebrate paleontology at KU, in 1983.[3]

Hesperornis, cited as evidence of the evolution of birds from dinosaurs, became a signature bird of Kansas. A recent book by Jill Hunting, also a combination of scientific biography and family memoir, recounts the discovery of the type specimen of *Hesperornis* by her great grandfather, Thomas H. Russell, a member of Marsh's Yale College (now Yale University) scientific expedition to Kansas in 1872.[4]

Both the flightless bird and the chalk itself became cultural references to Kansas. In the early days at the University of Kansas, chemistry professor E. H. S. Bailey was sponsor of the Kansas University Science Club. When the club gathered, they chanted, "Rah, Rah, Jayhawk!" By 1889, "Rock Chalk" replaced "Rah, Rah," referencing the limestone that exists in the Cretaceous-age bedrocks of the central and western parts of the state.[5]

At KU, the Rock Chalk Chant has become a well-known sound indicating imminent victory of the university's teams. A paleontological connection to the mythical bird, the Jayhawk, also evolved at KU. The school had received the Jayhawk designation as a nod to Kansas history. The term Jayhawker was applied to the guerilla fighters on the Kansas side during the Civil War, and eventually became a moniker for all Kansans.

It wasn't difficult to make the leap from *Hesperornis* to Jayhawk. In a fanciful essay for *Graduate Magazine,* alongside a drawing of a fossil *Hesperornis* with the caption "Ancestor of the Jayhawk," professor of geology Raymond C. Moore wrote, "[A] paleontologic field party from Yale University made the first discovery of ancestral Jayhawk bones in the cretaceous rocks of western Kansas. This bird was given the not unfitting name *Hesperornis regalis,* which means the 'kingly Western bird.' "[6]

Always curious about nature, we Bonner kids had lots of fun catching wild critters, such as lizards, bugs, toads, and turtles. The boys had a lark with Pharaoh cicadas, also called seventeen-year locusts. In other habitats, these prehistoric looking insects are dark and colorful, but in western Kansas they are grayish tan. The boys would hold the large insects safely in their closed hands, then move their fists around on their cheeks and chin, as if shaving. The locusts provided a shaver's buzzing noise. We also loved our adopted dogs, cats, hamsters and other mammals, so naturally, we came home from the beds one day with a warm-blooded pet.

During the 1965 visit of the Applegate family, while out in the chalk bluffs, Shelly and Dad heard a high, crying noise coming from under a rock ledge. They peered in to see an emaciated, crying raccoon baby. Everyone surmised her mother had been killed, so we brought the orphaned animal home to the house in Leoti. She was so little she didn't seem interested in solid food, so we used one of my plastic dolly bottles as a nursing bottle for the tiny mammal.

At Dad's suggestion, we named her Kitchie, a variation on the name of the mother wolf in Jack London's *White Fang.* Pop had read that book and London's *Call of the Wild* to us the previous winter. (One of his favorite lines from *White*

Fang, referring to the family of wolf pups, was, "With full bellies, bickering starts." He used it when we started fussing at each other.)

Kitchie was immensely entertaining. She was a social animal with her "family," digging bobby pins from the buns Mom and I wore, dancing on her hind legs for marshmallows, and finding treasures behind the couch. The wild side of her was never far away, however. She devoured all the guppies and goldfish we had in our fishbowl. Chris remembers that she even consumed the snails we had put in the fishbowl to help keep it clean.

We kept her in her own little den in the chicken house shed, on our side lot. Dana would take her to her den each night, placing her in a large open crate that had an old blanket in it. In the morning, one of us would bring her inside to play with the family. She enjoyed exploring the house and being with us much more than being alone. Sometimes she'd get a drink of water from the toilet, which prompted Dad to remark, "Daniel Boone left his hat on the stool!"

Kitchie was resourceful when it came to breaking and entering. No wonder bandits in the cartoons wear raccoon masks. Anything we left out was fair game. She consumed an entire jar of grape jelly that we left on the table when Mom took us to school one morning. When Mom came back to the house, Kitchie was sitting there, nonchalantly "fishing" in the depths of the jar. Another time Mom was horrified to see that Kitchie had flipped over a cake cover and plowed through the middle of a beautiful iced cake.

She stayed with us about nine months. The "call of the wild" kicked in as she matured. She started getting testier with us, biting me on the nose when I stared into her eyes too closely, and piercing Dad's thumb one time when he picked her up. She wouldn't stay in her den anymore, figuring out ways to escape the chicken house and regularly coming in and out of the kitchen screen door, which was never locked. She had the run of the neighborhood for several months, but one night she simply left.

Dad holds the pet raccoon, Kitchie, reaching for Chuck.

We assumed that she had gone to do raccoon things, have her own family, and live in the wild. Dad connected her life to the *White Fang* tale and explained it was only natural for her to want to be on her own.

One evening several months after she ran off, we had finished our homework and were sitting around visiting before bed, when we heard a soft banging noise. It sounded exactly like the door used to sound when Kitchie would open it. We sat stunned. We could hardly believe it when she waltzed right in and started walking around the group as if she remembered us. But chaos ensued when first Dad, then Chuck, then Dana, tried to pick her up to pet her. She bit them, one at a time. Mom and I didn't even try. Kitchie surveyed the living room, checked to see if there was any food around, then walked out through the kitchen door, never to be seen again. As we would recollect in later years, it was Dana who wittily asked, "Remember the time when Kitchie came in and *bit* us all goodbye?"

For decades, Dad didn't want to have a telephone. It was a strange objection, because his own father, Pappy, had one in the early days in his house north of the railroad tracks. But Dad claimed if we had a phone, people would be calling at all hours, asking the show times, what movie was playing, and other questions. He was probably right; a phone wasn't necessary for his movie theater customer service because exhibition times rarely varied, and posters in front of the theater always advertised the current feature. On the rare occasions when he needed to call someone, he went to the homes of kindly neighbors who let him make collect calls to people like George Sternberg.

In 1965, four years after the theater closed, Dad's anti-phone argument no longer held water. In the summer of that year, Clare Jane, Ray, and their kids were visiting Leoti. They planned it so that Clare Jane would get Mom and Dad out of the house. They meandered around town, even taking a trip to the cemetery, until Dad finally said to Clare Jane, "Have we done everything you want to do? I'm tired of running around." While they were out, a pre-installed phone line was activated and a black dial telephone appeared in the alcove off the living room.

When they returned, Clare Jane peered in the living room and saw the phone sitting in the alcove. She kept our parents in the kitchen by sitting down at the table and asking if there were any refreshments. Mom served them lemonade, and they kept chatting. Meanwhile, Ray, at a gas station two blocks north, called our new home phone number. Dad heard a phone ring. He thought it was a toy. It kept ringing. "Answer the phone!" he yelled down the hall where the kids were playing. Then he realized the ringing was coming from the alcove. He stared at Clare Jane. "Well, I'll be damned! What did you do?" He walked to the phone, picked it up, and said, "Bonner residence." Ray, laughing on the other end of the line, said, "We got you, Westerner!"

The year 1965 heralded another change. The theater building was finally ready to lease after being remodeled it into a large retail space. One of the most difficult tasks in the conversion was leveling the inclined floor of the theater. With the help of Leoti friends, Pop slowly "jacked up the floor" and stabilized it with drill stems from oil rigs. By autumn, the building was ready, and Mom called the grocery chain Independent Grocers Association (IGA). The large grocer wanted a franchise in Leoti, so they worked out a contract with my parents. The building's front was transformed with plate glass windows and automatic doors. Pop leased the building to IGA for one-hundred-twenty-five dollars a month.

After a couple of years, the IGA franchisee, Jim Curry, decided to declare his independence from IGA and bought the building from Dad. While Curry operated the grocery store, Chuck (briefly) and Dana worked as clerks and stockers. Curry sold the business in 1970 to Ralph Gropp, whose nephew, Bob Pepper, managed the store and renamed it Pepper's Grocery. He continued to employ Dana during his high school years.

Sadly, the theater building burned to the ground in 1973. It had had a long run, existing in its basic form for decades. In 1926, Wichita County purchased it after the tornado for use as an auditorium. O. W. Bonner bought it in 1937, and the Plaza Theater operated there for twenty years, from 1940 through 1960. It was a gloomy day when Dad realized the building was completely destroyed. "It's the end of an era," he said wistfully.

Nothing could have prepared us for the event that occurred on Friday, the thirteenth of June, 1967. At six in the morning, after returning from the bathroom, Mom lay back down in bed beside Dad and suffered a massive heart attack. She probably did not know the extent of her heart damage from rheumatic fever as a child, and she never talked about it to anyone. At the time of her death she was fifty years old, five of her children had left home, and the oldest three had started families of their own. Her hair was still black, long and soft. She wore it in a bun during the day and donned a hat when she was out in the wind.

When she began having trouble breathing, Dad tried to resuscitate her. She was unresponsive, and he began calling her name frantically. Dana and Chuck came from their room across the hall and tried to revive her with rudimentary CPR, which they had read about in *Life* magazine. I came to the door and asked if anyone had phoned the doctor. The Leoti doctor was not available, and by the time the doctor from Tribune arrived, Mom had been gone for about an hour. An ambulance took her away, and all of us went, stunned, back to our rooms.

When Dad came out of their bedroom, after composing himself, he sat down in the phone table alcove and began calling his children. He contacted a couple more people, including our Bonner cousin, Janet Dierks, who lived a few blocks west in Leoti. Janet came immediately to the house and took charge of calling all the rest of the extended family while Dad sat on the couch in the living room, numb.

Orv got the news in a delayed fashion because he was on a prehistoric bison dig near Meade, Kansas, with Dr. Claude Hibbard and a group from the University of Michigan.[7] Due to his proximity, though, he was the first to arrive in Leoti. The others came home as quickly as they could: Clare Jane from Arlington, South Dakota; Letty from Phoenix, Arizona; Steve from Oklahoma City, Oklahoma; and Chris from Kansas City, Missouri.

The headline of a story in the *Leoti Standard* two days later read, "Leoti Is Shocked by the Death of Margaret Bonner." Within the obituary, the newspaper editor wrote, "[She] passed into that Great Beyond within minutes after a fatal heart attack, about 6 a.m., at her home in Leoti. ... She was an excellent mother, an exceptionally good neighbor, and highly respected by all who knew her."

This crisis was the last thing any of us expected. I don't think Dad had ever envisioned a life without her. He was profoundly bereft after she died but put up a brave front for us. Like a tree blown raw by the wind, he vowed to raise the remaining three kids by himself, to finish the job she had started. He told me, "I'm not going to marry a farm woman just so you kids will have a mother. I taught Mom how to cook, so I will do the cooking now." That was very decent of him. No one could replace her anyway, so he didn't even try.

Mom's club commemorated her with quilts. First, the Mother's Study Club finished a quilt she had started of sunbonnet girls, whose dresses were made of material Mom had used in my sister's dresses over the years. After they finished this quilt, they gave it to me, and I still have it. Second, the same club finished a friendship quilt to honor her. It is viewable in the historic Washington-Ames House in Leoti, and its center block says it was made "in memory of Margaret Bonner" in 1968.

Several months after her death, Dad got in the habit of "tying one on" once a month or every six weeks. I came home from fifth grade one afternoon and saw Dad standing in the back yard, yelling. He swayed in the wind. I went over to him to ask him what was going on, and his answers were slurred. I thought to myself, "So *that's* what it looks like when someone is drunk!" It was the first time I had ever been exposed to inebriation. Drinking once again became a pressure valve for Dad, and we kids simply accepted it. I certainly never blamed him for it. I understood fully what the sobs coming through the bedroom wall were all about.

My memories of Mom center around her patience and her encouragement. She let Dad be Dad, never trying to change her husband, as some wives do. She extended this same freedom to all her children to develop their own personalities and follow their own paths. Even though I had her for only nine and a half years, the feelings I had for her were intense. I would have given my life for her without blinking. When her heart stopped, there was nothing any of us could have done. She was fifty and had lived a full life on the flatlands with the big rogue in her thrall.

Mom's was a quiet strength, both undergirding the family and giving it shape. When we lost her, it was as though the foundation rocked and the structure collapsed. Temporarily.

The older children came back in the summer of 1967 to help Dad, my brothers, and me adjust to the loss. Clare Jane, who was pregnant with her fourth child, a daughter she would name after Mom, stayed a week to help us figure out who would handle all the work Mom did in the home. Orv's wife, Jean, stayed with us most of that summer as well.

Fossil hunting was therapy. That summer, Dad made a huge discovery—a complete skeleton of the mosasaur *Platecarpus* ("flat wrist"). He spotted, in the middle of a knoll of hard yellow chalk, vertebrae that were located almost exactly in the middle of the swimming reptile's body. Squatting by the vertebrae, he dug off some chalk to the right, realizing that that side was headed toward the tail. He dug to the left and understood, with immense joy, that those vertebrae were headed toward the skull. As a skilled paleontologist, he could tell the fossil's orientation by the direction of the spinous and transverse processes (the bars of bone that connect the vertebrae to the ribs) and the angle of the ribs themselves.

The mosasaur quarry was six feet deep, and it took several trips back to the site to remove the overburden and prepare the fossil for collection. Because of the size of the specimen, Dad collected it in four pieces, dividing it at the skull and cervical vertebrae, the trunk or thoracic section, the lumbar vertebrae section, and a long, thin tail section. In a letter to the Askey family in September 1967, I wrote, "Daddy is almost finished boxing up the mosasaur. He is going to sell it to Los Angelas [*sic*]. He doesn't have to curate it." A page or two later, I reported, "Dad is dozing off. ...[He] has been feeling pretty tired because of the mosasaur."

This major discovery was one of the last fossils Dad sent to his friend Shelly Applegate, at a time when Shelly was preparing to leave the museum. Like a quiet predator lurking at the bottom of the Cretaceous seabed, the *Platecarpus* would lie in wait in a storage facility in Vernon, California, until scientists realized what a fantastic specimen it was. Over the years, the study of this singular specimen has added immensely to the knowledge of mosasaurs.

Dad knew that he needed to expand his museum connections beyond Kansas, Los Angeles, and Chicago. In 1966, before Mom's death, he had written Dr. Alfred M. Bailey at the Denver Museum of Natural History, a scientist he had met at the Denver SVP meeting in 1960.[8]

Alfred M. Bailey (1894–1978) was an esteemed ornithologist who served as director of the Denver Museum of Natural History for most of his working life, from 1936 until his retirement at age seventy-five in 1969. He was one of Dad's only museum contacts who was not principally a paleontologist. Much of Bailey's career focused on modern birds. Dad initiated the relationship with Bailey by donating some fossil bird bones (*Hesperornis*) to the museum in 1966. After that, he arranged to place with the Denver museum a complete *Pachyrhizodus*, *Gillicus*, and *Cimolichthys*, as well as a *Pteranodon* skull. Dad had agreed to bring the fossils to Denver on spec in May of 1967, the month before Mom died, so in the late summer he and the boys delivered the fossils to Bailey in Spiker.

As each son grew up and left home, Dad would turn to the next in line to become his fossil-hunting companion. I was lucky to be the last child because, even though I was a daughter, he tapped me to hunt with him. Dad did not treat me differently due to my gender. If anything, he began to understand women's rights much more in the years when I was growing up and encouraged me to leave Leoti and "be a career woman," perhaps even study paleontology in the future.

With no one to listen but me, Dad reminisced about everything he had experienced in his life. Wherever we were, he sang old songs, told jokes, or recited silly poems: "Old Mother Hubbard went to the cupboard to get her poor daughter a dress. But when she got there, the cupboard was bare, and so was her daughter, I guess!" In the fossil fields, setting the stage for lofty expectations, Dad would say, "Maybe we'll find something new to science today!"

One time, in the fall of 1967, Dad and I were hunting in the Big Place. It was an extremely windy day, even for western Kansas. I was looking around in some of the canyons with Dolli, who stuck close to me. I told Dad, "I think I'd like to climb to the top!" He replied, "Stay down in the lower bluffs. If you get up on top of those tall canyons, the wind could blow you off!"

I respected the admonition, for a while. But the lure of the wilderness and the challenge to my climbing abilities made me wonder what it would be like to go to the top and look down at the valley below. The ascent seemed like it was three hundred yards from the valley floor. It wasn't vertical but had several steep areas along some of the yawning draws that ran perpendicular to the earth. Why not? I thought. As I climbed upward, Dolli came with me. She seemed adventurous, as if she too wanted to check out new things she had never seen before.

About two thirds of the way up, we flushed out a great horned owl, and my heart stopped for a moment as the huge raptor drummed out toward us, then up and away. As we neared the top, Dolli ran ahead, and when I called for her, the wind drowned out my voice. That scared me a bit. But we were fine as we stood atop the high overlook, me singing at the top of my lungs, "Born free, as free as the wind blows / As free as the grass grows, born free to follow your heart."[9] I yelled as loud as I could, but the sound was a squeak in the wind. After a few minutes of powerful wind whipping my hair and clothes, my dog leaning into the gale to stay upright, I decided we had better make our way back down.

As we got about twenty feet down the bluff, I was on a hard yellow knoll and out of the wind when I saw a fossil, blue-gray against the golden bluff. The end of it looked like it might be a broken part of a *Niobrarateuthis bonneri*, the rounded part (rachis) of a fossil squid going into the chalk. This created a problem. Dad needed to know about this. I dug on it a little and determined it was angling down into the chalk.

I went back to Dad and confessed that I'd been up high but had found an interesting fossil. He was not happy that I disobeyed, but I fibbed and said I was following Dolli. I'm sure he saw the Bonner stubbornness writ large in me. To access my discovery, he said we should drive Spiker to the top of the bluff and then climb down to the fossil. We went back to the car, and he navigated his way around the canyons so that he could park above the knoll.

When he saw the fossil, he immediately said, "It's the round part of a squid, and it looks like it might be there!" His mood changed from irritated to excited. The rock was so hard, he didn't need any plaster and decided to dig it out in a rectangular block. What happened next astounded us both. Accustomed to the length and roundness of *Niobrarateuthis bonneri*, he dug around the sides of the fossil and trenched deep enough to lever it out. But when he pried it out and turned it over, the pointed end of it protruded from the bottom of the block by a good inch. "God!" he said. "This is very different. This is pointy instead of round." He said it was lucky the tip hadn't broken off. It was like someone had heaved a large spear from above, and it lodged in the sediment, waiting to be discovered millions of years later.

Dad took the fossil to Myrl Walker at the Fort Hays museum. In the introduction to the paper describing *Enchoteuthis* as a new genus, university scientists Halsey W. Miller and Walker summarized:

> Fossil coleoids [the group of cephalopods containing octopi, squid, and cuttlefish] from the Niobrara Formation of Kansas are rare, although several well preserved specimens have been found. Logan (1898) described a member of the Palaeololiginidae *Tusoteuthis longa,* and Miller (1957) described a Kaelenid [*sic*],[10] *Niobrarateuthis bonneri.* In the fall of 1967 Melanie Bonner, daughter of M. C. Bonner of Leoti, Kansas, discovered a new Palaeololiginid that is herein described as *Enchoteuthis melanae.*[11]

The "teuthis" portion of the genus name means squid. Dad's type specimen of 1957 meant "Niobrara squid" and mine meant "spear squid." The primary difference between *Niobrarateuthis bonneri* and *Enchoteuthis melanae* was that the rachis of the first was oval-shaped and that of the second was spear-shaped. (Photographs of both fossils appear on page 159.) But in a few years, much to Dad's disappointment, paleontologists began challenging Miller's designation of these squids as separate genuses from *Tusoteuthis longa,* which was also pointed.

Some scholars said that naming an invertebrate by only the gladius and rachis features was inadequate. Also, both the Miller descriptions were made by analyzing the underside rather than the dorsal side of the gladius.[12] Those who support the assertion that the squids are all *Tusoteuthis longa* believe that even though *Niobrarateuthis* is smooth and round compared to the other spear-shaped rachises, it could have eroded to be that shape. Dad maintained that his squid was collected and prepared exactly as it was found, in an oval shape.

To make things even more confusing, *Enchoteuthis melanae* is currently accepted by some paleontologists as a separate genus, whereas *Niobrarateuthis bonneri* is not.[13] But overall, most fossil-squid experts currently believe that all hitherto described squids of the Niobrara chalk are in the taxon *Tusoteuthis longa,* the first name established for these invertebrate animals.[14] It's likely that neither of the Miller-described teuthids found by Bonners will hold up, having both become junior synonyms.[15]

Myrl V. Walker (1903–1985) started collecting fossils while a student at Fort Hays State College under the tutelage of George Sternberg. He earned his master's in vertebrate paleontology at the University of Kansas in Lawrence and later became assistant professor of geology and director of the museum at Fort Hays from 1955 to 1970. Sternberg and Walker were both advisors to my brother Orville, who got his master's degree from Fort Hays and wrote his thesis on Dad's plesiosaur and a *Nyctosaurus* flying reptile found by Sternberg. Both scientists recommended Orv for a position that opened up in the Dyche Museum at the University of Kansas in Lawrence.[16]

I remember Walker as a kind soul. He checked in on Dad often after Mom's death. Knowing that Chuck was graduating from high school the next year and was interested in attending Fort Hays State University, Walker asked if Chuck would like a work-study position in the museum. When Dana went to Hays three years later, Walker repeated his generosity and ensured Dana had a museum job as well.

The son who had made paleontology his career was hired for a preparator position in the vertebrate paleontology department at the University of Kansas in 1968. After Orv got his job there, we younger members of the family started visiting the Dyche Museum religiously, just as we had the Fort Hays museum.

PART IV - LEGACY

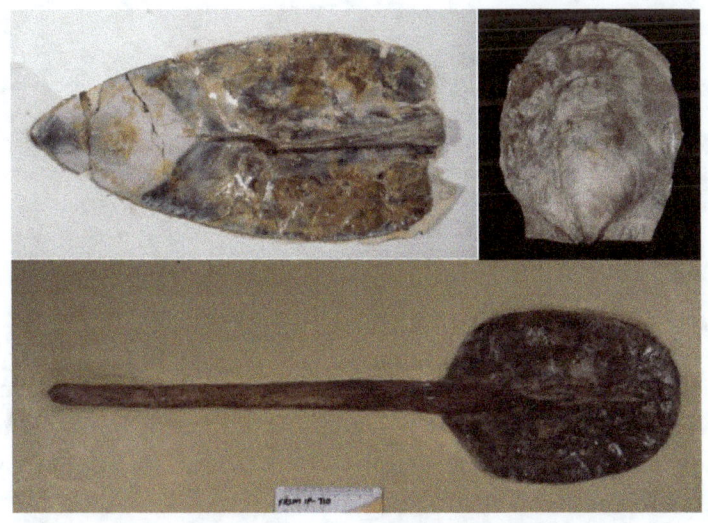

Previous page: Three invertebrate fossils (two squids and a bivalve mollusk) named for Bonners, clockwise from top left: *Enchoteuthis melanae, Pecten bonneri,* and *Niobrarateuthis bonneri.* Images are not to scale. Photos courtesy of the Sternberg Museum of Natural History.

My Muse
1980s

As my parents' last child, I enjoyed plenty of solo time with them and felt their happiness when they explored western Kansas. By the time they had me, they were as benevolent as grandparents, or perhaps they were simply worn out. Indeed, they had become grandparents the year I was born.

When they investigated outcroppings, abandoned houses, and historic sites together, they took me along. At four years old, I was extremely jealous of the attention my father paid my mother, and he played that to the hilt, sending me into a tizzy by making loud kissing noises while walking toward her, arms extended like Frankenstein's monster. I would scream, "No!" and run to protect her from his advances like a sheepdog guarding a lamb. She would chuckle at the contrived drama or say, "Oh, Marion!" I remember distinctly putting a grubby little hand between their lips as he zoomed in for a big smooch, their lips smacking on both sides of my hand.

Dad and Mom celebrated their twenty-fifth wedding anniversary in September of 1960. Clare Jane made a cake for the occasion and decorated it with silver bead candies, since it was their silver anniversary. I stared at that cake and couldn't resist tasting those little decorations, and that cake grab is documented in a family snapshot. Mom and Dad weren't too worried about that, as they kissed over the cake for their anniversary photo. Obsessed with the candy decorations, this was one time I didn't object to him kissing "*my* Momma" passionately.

One fall morning after Chuck, Dana, and Chris had gone off to school, Dad said to me, "Mellie, would you like to go with me and Mom to Gove to look through some old houses?" I replied vehemently, "No!" "Why not?" he asked. "'Cause you'll just tickle her ... all the way!" They had a good laugh at this, and of course, Dad slapped a seed cap on his by-then-bald head, Mom donned her straw boater hat, and we all went to explore some ghost towns.

As protective as I was of Mom, I was close to Dad too. He pretended to involve me in important work when he was collecting fossils, giving me little tasks to help the day move along. During the summer months in the fossil-bed sun, my long locks turned the same sandy blond Dad's had been when he was young, and the winter months brought the color back to medium brown. Mom often put my hair in a bun like hers to keep it under control.

Mother had never missed a day of church unless we were out of town or she was having a baby. Her influence was strong, but the Old Man's was stronger. To the townspeople, despite the fact we had perfect attendance at Sunday School in the Presbyterian Church, we must have seemed like feral children. Sunday was the one day of the week we all cleaned up and exhibited our best behavior. But that civilizing effect would fade during the week as we ran around Leoti, and by Saturday, when it was time to hit the fossil beds, we were ready to breathe the smell of chalk and dry grass and be filled with the hope of hitting the jackpot.

We were a little like the indigenous wildlife of the fossil beds ourselves, communing with the critters. We captured countless lizards and rescued the baby raccoon, but for the most part we could only observe many other kinds of animals. *Roadside Kansas* summarizes the visible fauna of the fossil beds and pastures:

> The native vegetation and availability of surface water make the Smoky Hill River valley an ideal habitat for large herds of mule deer and pronghorn antelope, which were once widespread over the western two-thirds of Kansas. Pronghorns are not true antelope, but they have characteristics of deer and goats. They are but one of the native Plains animals that were misnamed by the pioneers. Prairie dogs are actually squirrels; jack rabbits are actually hares; buffalo are bison; elk are really wapiti; prairie chickens are grouse, and horned toads are actually lizards.[1]

One time when my sister Letty and her family were visiting in the fall of 1966, the stifling heat of summer had abated, and the animals were active. When we exited the fossil beds, we saw a family of kit foxes—a mother and three pups—near the gatepost. Kit foxes do not have the red coloring and big plumed tail of other foxes. They blend into the tan grasslands and look a bit like small coyotes, but downier. Their black-tipped tails, extremely large ears, sharp muzzles, and whimpering noises were adorable, but Mama fox wouldn't let us get close to them, protectively skittering off, yipping for them to follow. I couldn't help making a mental comparison between the two mamas, both vixens, Letty with her brood of three and the defensive fox mother.

On a different outing, we watched a family of striped skunks parade across the road. These animals, with their black-and-white markings, stood in pronounced contrast to the dull brown landscape. Dad stopped Spiker for us to observe them but also because hitting a skunk would be disastrous to the animal as well as to our noses. These mammals were also endearing, their white striped tails hiked high in the air as they made their way single-file into the ditch.

On a fall day in 1968, Chuck and Dana hunted ducks while Dad and I hunted fossils. I'm holding our miniature poodle, Dolli.

Birds were plentiful as well. I liked leafing through Dad's *Field Guide to Western Birds* in order to identify them. Most recognizable was the western meadowlark, songbird of the plains and state bird of Kansas, which perched on fenceposts and filled the air with its piercing, distinctive melody. Hawks were plentiful and of many varieties, catching thermals and soaring above, keeping a fierce eye out for prey below. Small birds such as sparrows, finches, and wrens made an appearance, but among the perpendicular bluffs, the easiest bird to spot was the cliff swallow. They darted in and out of their mud-daubed homes and were quiet during the day but could be heard squeaking and chittering around dusk.

After Mom died, Dad would go up to the fossil beds by himself, working through his grief, and I imagine him talking to her as he was searching for and working on specimens. He always took our family pet with him and talked to her as well.

One autumn day, Dad was scratching on a fossil in some hard, waxy chalk with his small tool, which made a cheeping noise as he worked. This noise called up a coyote. Dolli went on alert, and so did Pop, both of them gazing up at the coyote that had come over the bluff and was staring down at them. The coyote saw that the cheeping noise was being made by a man, not small game, so he turned and lit out at a fast run. Dolli chased him, and the two canines ran out over the top of the pasture. Pop watched helplessly as the coyote, at first galloping vigorously, started slowing down to a trot, then started to turn and confront the miniature poodle. Dad yelled for Dolli frantically, and she finally paid attention and ran back to him, escaping what could have been her own *White Fang* encounter. "I pert' near lost her—had a hell of a time getting her to come back," he told us. He marveled at how the coyote seemed at first curious, then skittish, then emboldened and "crafty."

Chuck recalls, "One of the neatest times we experienced nature was when Dad, Dana, and I were collecting on the north side of the Big Place in Spiker." Dad and my brothers had decided to stay overnight on this trip, so they had their dinner of potato chips, sodas, and peanut-butter-and-jelly sandwiches and were watching the summer sunset. "We saw an entire herd of deer leap up and run to the summit of the Big Place. The sun was setting, and the deer got to the top, stopped, and just stood there looking at us."

That night they camped, sleeping in the front seat and the back of Spiker, and heard coyotes yipping all night.

An evocative place for me during my undergraduate years was the iconic building that housed the museum on the KU campus where my brother Orv worked. It is famous for dioramas of North American animals, many of them collected by Lewis Lindsay Dyche. Dyche Hall is a limestone Romanesque edifice on Jayhawk Boulevard that resembles a French medieval church, but is in fact a church to science. The tile floor in its entryway is famous for its visual depiction of evolution in successive triangles, starting from an ancient shark on the bottom and ending with a human being at the top.[2] On the building's exterior, etched in stone around the top windows of the structure, are the names of groundbreaking scientists: Huxley, Darwin, Audubon, Gray, Agassiz, and Cope.

Two men whose names were not etched in stone but who made the museum possible were Francis Huntington Snow and E. Raymond Hall. Snow was a member of the three original KU faculty, and Hall was director of the museum from 1944 through 1967.[3] Snow and Hall both persuaded the Kansas Legislature to fund the construction and expansion of the building. Before his retirement, Hall approved the hiring of my brother Orville to be a preparator in the vertebrate paleontology department.

In my freshman year at the University of Kansas, I would have lived at the museum had I been allowed to. It was familiar—its framed and mounted fossils reminding me of the Fort Hays museum and my home. Missing Dad, I wrote to him often, talking about my museum job. I had obtained a work-study position there, thanks to Orv. Dad sent me letters and called at least once a week during my freshman year. One time he sent a letter that had a photo in it. On the back, he had written a description in his characteristic cursive: "This was the last photo of Mom taken in the fossil beds at the turtle quarry." The corner of it is torn from when I hung it on a bulletin board, but I have kept it all these years. Dad always updated me on his discoveries, whether on the phone or in letters. I loved seeing my name in his large scrawl on the letters I received at school.

The last photo of my parents, taken in the fossil beds a few months before Mom's death in 1967.

Orvie's preparator job in the museum so impressed me that I briefly considered majoring in geology, hoping to attain a doctorate and become a museum paleontologist. But it didn't take much analysis to understand that those jobs were so scarce, getting one would be nothing short of a miracle. Plus, I was completely intimidated by the fact that I would need to take scientific calculus. That scotched it for me. I considered education and pre-law before finally deciding simply to take courses that interested me. I enrolled in a full semester of English courses, including American poetry, Shakespeare, and other topics near and dear to my father's heart. My favorite undergraduate professors were David Bergeron (Shakespeare), Beverly Boyd (Chaucer), and Vic Contoski (American literature and poetry).

Shortly before graduating, I decided to pursue literature further, so I took the Graduate Record Exam and received a spot in the English department as a graduate teaching assistant. I would teach—with tons of professorial and peer guidance—freshman composition and literature courses while working toward my master's degree in English. University of Kansas students working toward their master's had the option of taking traditional literature courses or following a creative writing track. I opted for the former.

When I was in graduate school, I met many people who had an influence on my education, but two stand out: Victorian literature professor Roy Gridley, who would become my thesis adviser, and my officemate and fellow teaching assistant Denise Low, who, decades later, would become poet laureate of Kansas. Both were native Kansans and kindred spirits. And both of them reminded me of my parents because they were refreshing, intellectually robust Kansans.

Studying English and American writers was stimulating, but I knew I was not like other graduate students, and I was averse to writing a detailed scholarly analysis of someone like Johnathan Swift or Charles Dickens. Unashamedly, I wanted to learn more about the land I grew up in. Regional literature was becoming more accepted in academia, and one of the new courses offered in the English department was taught by Professor Gridley, a native of Oakley and the only Kansan on the English department faculty at that time. The course was Great Plains Literature, and I knew this was my opportunity to learn about literary references to the plains and to flesh out what I had learned from Dad about the historic and prehistoric background of our home turf.

I should note, even in major universities of the Great Plains, the study of regional literature is a side gig. For example, Gridley's primary area of scholarship was Robert Browning, and Linda Ray Pratt at the University of Nebraska, author of *Great Plains Literature,* is an expert on Matthew Arnold. An example of how Gridley blended Great Plains studies with traditional English scholarship was his essay, "Some Versions of the Primitive and the Pastoral on the Great Plains of America," which was included in *Survivals of Pastoral,* edited by fellow English professor Richard F. Hardin, in 1979.[4]

A few weeks into the Great Plains course, my mind started bending toward potential thesis topics. The obscure book, *Buffalo Land,* by W. E. Webb, seemed like a good place to start, and I happily realized that I could share with Dad what I discovered in graduate school about plains literature. As the course neared completion, I wrote the required term paper on Webb's book. I saw in it multiple literary allusions and cultural references, such as "Fire on the Plains as It Is vs. Fire on the Plains in Novels" or "A Pickwickian Blunder." I floated the idea of doing an annotated edition of the work for my thesis. I could reproduce the book, provide notes for the text, and write an introduction. For the thesis's abstract and introduction, Professor Gridley encouraged me to place it in the American literary canon and examine its cultural significance. At the time, I wasn't focused on the fossil information in the book provided by Cope.

Other graduate students and professors joked about the length of my thesis: Volume I reproduced the large book in spreads and added annotation numbers, whereas Volume II (also more than two hundred pages) provided an analysis of where the book fit in the plains travel narrative genre and included notes to the reproduced volume. My notes enlarged with incessant commentary on Webb's racist and cultural stereotypes, which were shocking to me in the 1980s. He horribly lampooned Native Americans, Irish and Mexican immigrants, and the "genus Texan," a rough, unprincipled cowboy. His mockery of the defeated Indians around Hays City, who were struggling with alcohol and loss, was particularly galling. I'm sure my overwrought thesis, which now seems like a fossil itself, lies buried in KU's Watson Library thesis archives taking up three inches of valuable space.

Sadly, Professor Gridley died in 2015. A Rhodes scholar, he received his bachelor's from the University of Kansas in 1957, his master's from Brown University in 1959, and his doctorate from the University of Illinois in 1964. He wrote a literary biography of Browning in 1972, and *The Brownings and France: A Chronicle and Commentary,* was published in 1982. He taught English at KU from 1964 through 1997, and beginning in 1979 frequently spent summers in China assisting his beloved wife, Marilyn, an art historian and fellow western Kansan, with her research. Like my parents, they were another intellectually curious pair from Kansas.

My first officemate in Wescoe Hall had already attained her master's in English and was employed as a full-time lecturer in KU's English department. At the time I met her, Denise Low was an experienced instructor and mother of two boys who was editing the college literary magazine, *Cottonwood Review.* Originally from Emporia, Kansas, Denise greatly encouraged me after I joined her in our office. By the time I had one semester under my belt, I was attempting to write poetry and short fiction as well as working toward the traditional master's degree in English. After I had shared the office with her for one year, helping her with the production of the journal, Denise generously handed over the reins of *Cottonwood Review* to me. I managed it for a year but felt like a pretender as a poetry and fiction editor. I was more comfortable with the book production process than with creative writing and editing. Erleen Christensen became my office mate after Denise, and I gratefully passed *Cottonwood Review* over to her creative hands.

After her teaching stint at KU, Denise Low earned a doctorate in English at the University of Kansas and an MFA in creative writing from Wichita State University. She became a professor and administrator at Haskell Indian Nations University in Lawrence for twenty-seven years. She was named the second Kansas poet laureate in 2007. In her career as a creative writer and editor, she has published more than twenty books of poetry and essays and founded Mammoth Publications with her husband, Thomas Pecore Weso. In addition to poetry volumes, she has authored or edited more than a dozen books.

I was delighted to catch up with Denise after decades apart, most of the time for me in Texas. Her voice sounded exactly the same as it did in the 1970s. She has written many poems influenced by the Kansas landscape and fossils. What touched me most about her work, and was most coincidental during my writing of this book, is the beautiful memoir she penned about her lost Native American ancestry in Kansas as part of the American Indian Lives series, produced by the University of Nebraska Press. In it, as in many of her works, the Kansas landscape is a presence, an influence, a character as she traverses the state, digging at the roots of her ancestry. She writes that in the grasslands, "the sky can fill a full circle. Two time frames, past and present, comingle. In remote places I explore where my grandfather once lived. After dozens of trips west into the plains, I came to identify Grandfather with the land itself. The sun rises on the rough-etched outline of his form; it sinks into the western horizon with blazes of his fiery breath."[5]

I now realize I subconsciously chose a thesis topic that kept me close to Dad. Driving out to western Kansas to visit him frequently during my graduate school years created a mental shift, a break from my brief acquaintance with academia that soothed and centered me. I chose analysis of Webb as a labor of love because I wanted to closely examine the things that interested my father. During those years, he and I moved past the parent-child relationship and developed something akin to adult best-friendship. My in-depth excavation of *Buffalo Land* and the literature of the plains, very like a fossil, was a way to justify his and my existence. We are here. We mean something.

As Denise Low, Roy Gridley, and many other Kansas poets and authors have expressed in their works, the Great Plains is a character in our lives. It definitely was for Dad and me, together and separately. And the layer under the Great Plains, the fossil beds, was for members of the Bonner family, a signature presence.

The smell of chalk unites us with the past, and the place that underpins and overreaches us. When we are on the grasslands and in the eroded bluffs, we are in two time zones simultaneously.

Having been brought up by a self-taught scholar, I was motivated to share my scholarship with Pop. From a psychological standpoint, I have spent my whole life trying to make my father proud. That driving need was rooted in me from childhood. Also, it occurred to me as I wrote this book, that subconsciously I was trying to reconnect with my mother as well, the woman who sent Dad down this historical, literary, and scientific journey with Webb's book, *Buffalo Land*. Both of my parents loved language. Dad made us all groan with his incessant punning, but he truly was an artist with words. He was pleased I was an English major, calling me a "word woman" when I added new terms to his vocabulary. During these years, with my increased awareness of American literature, I viewed him as a figure akin to James Fenimore Cooper's elderly Natty Bumppo in *The Prairie*.

I shared with him poetic discoveries, such as stanzas from Tennyson's *In Memoriam,* that touched on our experience. Lines that moved us were well-known: "'Tis better to have loved and lost / Than never to have loved at all" (canto 27) and obscure: "Tears of the widower, when he sees / A late-lost form that sleep reveals / And moves his doubtful arms, and feels / Her place is empty, fall like these" (canto 13). But beyond emotional connection, Pop most enjoyed (and repeated often) Tennyson's stanzas about geologic time that I used as an epigraph at the beginning of this book just before the Table of Contents.

In my twenties, I summarized the influence of my upbringing and geography in a poem, "The Plains as Restorative." This poem was one of the few chosen for publication during my Kansas years. Denise Low (very graciously) included it in *Confluence: Contemporary Kansas Poetry,* which she edited and published in 1983.[6] The title comes from a popular view of the plains by early travelers. As I read it now, I recognize how clearly Dad was my muse. Literature and scientific study of the plains—a simple and beautiful place no one had heard of—held their own kind of spell over me.

Each section of the poem has a different purpose. Part I is historic, Part II is about the loss of Mom and the embrace of my family, Part III is about Dad in the (then) present, and Part IV is about personal transformation. It hints at moving out of Kansas while a part of me stays there, like a trace fossil.

The Plains as Restorative

I

Josiah Gregg, Santa Fe trader, takes eight trips
across the prairie ocean.
A Missouri doctor prescribes dry plains air
for his consumptive lungs.
Modest, reclusive Josiah Gregg finds knowledge
of plains flora, fauna, phenomena
and becomes addicted to his cure.
He noons at the Elysian vale
of the Cimarron,
breathes freely, sweats happily.
There are no flies this far West.

II

Long ago, mother died.
So I roamed the breathing abdomen of a continent.
Ridges of loess left by Pleistocene dust storms
rippled over Nebraska.
In eastern South Dakota
ridges flattened into rolling green.
Black plowed land undulated,
empty as the day it was born.
The northern herd seemed to rise out of black loam;
dollar-green farm crops blinked to soft, mossy hide.
I thought of that sister waiting to hold me,
and I was nearly home.

III

He's here somewhere breathing in these fossil beds,
his persistent eyes scanning yellow chalk walls
and blue gully bottoms.
Sweat bees hum around my ears.
I trip on brittle grass and fall forward,
catching myself on a chalk wall
that absorbs hands' sweat like a slow sponge.
An inhalation dries my lungs.

Arid space clarifies each dirt-capped knoll.
Small heat waves skirt the whole horizon,
making low-lying hills ripple.
Now, as brown lizards dart and grasshoppers click
away from my moving feet,
I hear my father's pick striking chalk.

IV
You have been captured.
Drops of sweat run like greasy ants down your neck.
Lie still, with your ear to the ground,
your head and chest pressed into sand.
Four ropes pull you taut.
The sound of blood moves in your ears
like a slippered, shuffling old man.
Unstake your spirit.
Watch your dead body burn into desert sand.
Weightless now, stare with rising eyes at high cholla
until the moment when thin clouds stand quiet
and cactus begins to move slowly west.
Leave flesh to dissolve and bones to bleach.
That slice of you stays,
while your journey spreads, horizonward, into dreams.

Institutional Knowledge
Late 1960s–70s

OVER THE YEARS, AS Dad got older and lived alone, he phoned all of us often. When he spoke with Orv, paleontology was their common language. "I went over to Shields and wet down 'Slimmie' the *Saurodon*," Dad said. For Slimmie, Dad used a little water to help remove chalk but he also discussed how he was strengthening the fossil. Orv remembers, "George [Sternberg] showed Dad how to harden the fragile fish bones with gum acacia, then later, with Glyptal varnish, which was an insulator of circuit boards. They also used Campbell's cement, which was originally an airplane wing adhesive."

Having instilled in Orv the love of hunting and collecting fossils, Dad vicariously—and predictably—enjoyed Orv's studies in geology and paleontology when he went to college, and each move he made in his career influenced Pop and the rest of the family.

Back when he was a student at Fort Hays, Orville worked in the museum with both George and Myrl Walker and mastered the Sternberg methods of collection and preparation. Orv met his wife, Jean, in college, and they were married in December 1961. In 1964, she typed his master's thesis on Dad's *Dolichorhynchops* plesiosaur (then called *Trinacromerum*) and a *Nyctosaurus* flying reptile found by Sternberg.[1] In the paper, Orv proffered a new species name, *Nyctosaurus sternbergi*, after Sternberg.

After FHSU, Orv worked at the University of Michigan at Ann Arbor in the fall of 1967 and spring of 1968 under the tutelage of Dr. Claude W. Hibbard (1905–1973), another respected Kansas paleontologist. "Hibbie" was born in Toronto, Kansas, and graduated from the University of Kansas in 1933. He earned his doctorate from the University of Michigan and was curator of vertebrate paleontology at the University of Michigan Museum of Paleontology for twenty-seven years.[2]

During Orv's time in Ann Arbor, he received word of the opening in the vertebrate paleontology department at KU. He applied for the position and received letters of recommendation from Hibbard, Sternberg, and Walker.

It was a plum job. To be hired as a full-time museum paleontologist at the University was a "feather in his cap," as Dad put it. Positions like these were rare across the U.S. Nowadays, to get such a spot, one must have a doctorate in vertebrate paleontology. As it was, Orville brought something as valuable as a higher degree: years of practical experience, which he shared with many graduate students over the years.

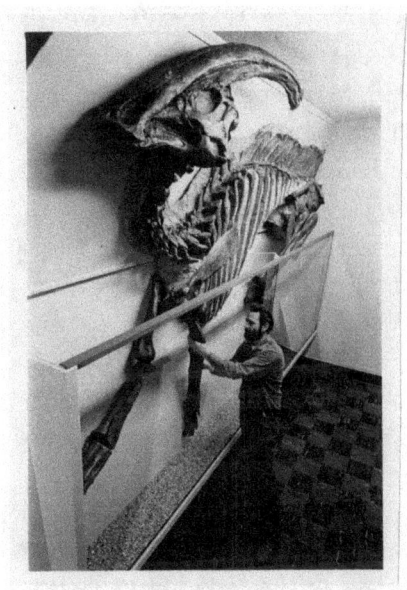

Orv was hired to be a preparator after Russell R. Camp, who served in the role from 1945 to 1967, passed away. Orv filled the preparator role until his own retirement. From April 1968 through September 1997, Orville worked with four paleontology curators: Dr. Ted Eaton, Dr. Craig Black, Dr. Larry D. Martin, and Dr. Hans-Peter Schultze. Eaton was curator until 1978, Black from 1970 to 1972, Martin from 1972 to

Preparator Orville Bonner adjusts a cast of Parasaurolophus on exhibit in the KU Museum of Natural History, 1977. Photo courtesy of Kenneth Spencer Research Library, University of Kansas.

2012, and Schultze from 1978 to the present, now serving as curator emeritus.[3]

Dyche Hall was the single most influential building during my years at KU. The museum itself has a fascinating history. Its founder, Francis H. Snow, was a naturalist specializing in botany, zoology, entomology, and climatology. He collected countless specimens for the museum, which started in the University of Kansas's North College as a "Cabinet of Natural History."[4] Before installation in Dyche Hall, the "cabinet" was also housed in Old Fraser Hall and Snow Hall. The current Romanesque museum building was named for adventurer Lewis Lindsay Dyche, and the large, detailed wildlife panorama of modern animals that Dyche prepared for the 1893 Chicago World's Fair is still on display. In addition to mentoring Dyche, Snow was instrumental in bringing the well-known paleontologist Samuel W. Williston to the University of Kansas in Lawrence. Before agreeing to become chancellor of the university, Snow insisted they hire a world-class scientist, and he wanted Williston.[5]

Samuel Wendell Williston (1851–1918) was involved in collecting fossils from Kansas for O. C. Marsh of "Bone Wars" fame. He later collected fossils with Cope, then "became a world authority on fossil reptiles."[6] Williston was acquainted with Benjamin F. Mudge from his days growing up in Manhattan, Kansas, where Mudge was a professor of natural history at Kansas State University. He worked with Mudge collecting large dinosaurs in Wyoming. A polymath, Williston attained degrees in medicine and entomology and headed up both the University of Kansas School of Medicine and the museum's paleontology department for twelve years. In 1902, Williston left Kansas to direct the department of paleontology at the University of Chicago. In his biography of Charles D. Bunker, another Dyche Museum man, Chuck Warner writes, "Today the legacy of Samuel Wendell Williston at the University of Kansas remains significant. Among his major contributions is the wealth of fossil specimens he collected for the KU museum from the Cretaceous Niobrara Chalk. ... But he also left something else, lesser known but likely just as important: his commitment to training young scientists."[7]

Describing a visit with Orv in the KU museum in the 1994 book *Planet Ocean*, Brad Matsen wrote, "Marion's oldest son, Orville, studied with George Sternberg at Fort Hays State and became a collector and preparator at the University of Kansas, where, over his desk in the museum basement, he has a sixteen-by-twenty-inch, framed photograph of E. D. Cope."[8] That was a handy way to link paleontologists from the early days to the present. Orv says, however, that the framed photograph was of the venerable Williston, not Cope. Either way, the historical thread works.

When you have a Dad who talks about bones all the time, it is natural to want to identify bones wherever you find them. When our mother made chicken dinners for the family, without fail, Chuck and Dana would identify the bones on our plates, such as the humerus, radius, and ulna from the wing I had just eaten. An eye for bird bones paid off for Dana on two back-to-back fossil bed jaunts.

Whereas bones of the flightless diving bird *Hesperornis* had been found by Dad, Mom, Orv, and Chuck, sightings of the smaller bird of the Cretaceous, the *Ichthyornis*, were much rarer for my family.

In early spring of 1964, when Orv was at FHSU, Dad, Orv, Chuck, and Dana decided to hunt the Wishbone, a relatively small outcrop in Logan County with two long, connected ravines that looked like a wishbone on the map. On the left side of one of the canyons, while Pop and Orv were working on a *Pachyrhizodus*, Dana, then eleven, went roaming. Circling some of the small knolls, he looked down on a shaly overhang and spotted a strange white bone. It looked odd yet also familiar. "It was very weird—I wasn't even sure if it was a fossil at first," he recalls. "I looked all around and below the bone but didn't see any other bones like it." He collected the single bone and took it over to show Dad.

Dad sat up abruptly, filled with excitement. "Where did you find this? Show us!" Dana took the two back over to the knoll where he had picked it up. They also scraped the area and found no other elements. Dad told him it was a bone of a rare bird in the Smoky Hill Chalk, an *Ichthyornis*. The *Ichthyornis* ("fish bird") was a toothed bird that resembled a gull or tern. It most likely flew like modern birds. The fossil hunters carried their enthusiasm back to Leoti, where they told Mom, Jean, and me about the bone. It was the distal end of the tarsometatarsus, a limb bone that would have been connected to the bird's foot.

The next day, when Pop and the family went back to finish collecting the *Pachyrhizodus,* something inconceivable happened. Dana continued hunting in the canyon where he found the *Ichthyornis* bone, but this time he focused on the opposite side of the draw. He spotted, in the side of a knoll, a thin bone running parallel with the chalk striations. He didn't dig on it but called Orv and Dad over. Unbelievably, he had found the radius and ulna of a second *Ichthyornis,* which looked very much like the chicken bones on my plate.

Orv took the three bird bones back to Fort Hays, where they are now in the museum's collections. Dana said of the three bones, "It was a very unusual experience. To find those two distinctly separate parts of an *Ichthyornis* on two back-to-back days was just incredible." For a moment, Dana said he felt like a real "bird man." But, he added, "I had never found any *Ichthyornis* bones before ... or since."

Pop later found some scattered *Ichthyornis* elements that ended up in the Los Angeles County Museum. He sent a tarsometatarsus and a partial skeleton that included a mandible in the 1970s. One of the most complete specimens is at the Sternberg Museum, discovered by J. D. Stewart, a paleontologist from Plainville, Kansas, near Hays. The jaws of Stewart's *Ichthyornis* held teeth, which added important data to the study of this genus. In 2015, Chuck delivered a significant skull of the elusive bird to the American Museum of Natural History in New York. It later became the subject of an avian skull study headed by Dr. Julia Clarke of the University of Texas at Austin.[9]

The *Pachyrhizodus* material found and collected by Orville that weekend also went to Fort Hays. The *Pachyrhizodus* (one of the most fun names to say) was a medium-sized (two- to eight-feet long) ray-finned predatory fish that had conical teeth like a mosasaur's.

After Shelly Applegate left Los Angeles, Dad wondered if he would be able to stay connected to the LACM. Fortunately, after Applegate's departure, the museum hired Kansas-born and -educated paleontologist J. D. Stewart. Stewart earned his master's degree from the University of Kansas in 1979. He studied Pleistocene microfauna under Dr. Larry D. Martin (no relation to Handel T. Martin). Martin, a Nebraska native, was widely known as a bird expert but was also knowledgeable of most vertebrate fauna.[10] In 1984, Stewart attained his doctorate under Dr. Ed Wiley in the KU ichthyology department, researching Cretaceous fishes found in Inoceramids (large clams).

The largest of the Inoceramids of the Kansas chalk was *Inoceramus platinus,* which could grow to four feet wide. The small fish that sometimes fossilize inside *Inoceramus* ("strong pot") were similar to herrings and small perch. The giant clams provide a source of small fossil fish that otherwise would not have been preserved.

As a graduate student, Stewart worked in the field with Orville often. Once, he and Orv were standing on a precipitous spire north of Castle Rock. Orv was clearing away chalk from around a large *Inoceramus.* J. D. stood on a ledge, watching him unearth the fossil. "All of a sudden, the ledge broke off," J. D. remembers. "I didn't know what to do, but instinct kicked in and I started running. The only way I wasn't going to be severely injured was to move my feet as quickly as possible down the steep slope." Running straight down the cliff, when he got near the bottom, he fell and got a few scrapes. Later, Orv commented on how well he handled himself and added, "I thought you were a goner!"

It was during these visits to western Kansas, when the KU hunters were relaxing after a long day in the chalk, that Dad came over to visit his son and the other paleontologists. He regaled Larry Martin, J. D., Orville, and the other graduate students with tales from the old days. J. D. shared his memories of Dad with me while I was researching this book. He recalled the Plute Clayton tornado tale ("Thousands killed, and many injured …") as well as farming stories involving "one-way combines, steam engines, and Rumely tractors."

Arriving in Los Angeles in 1986, Stewart knew a great deal about Kansas fossils and would be instrumental in understanding the specimens in the years to come, making an important rediscovery of a major Bonner fossil in 1988 and acquiring a now-famous plesiosaur. Pop was thrilled there was a Kansan at the museum. This prolonged the California connection that began with Shelly Applegate.

Dad had basic drawing skills, which he had used to illustrate fossils in his high school term paper, "Cretaceous Fishes." Mom was a talented artist, always encouraging us to paint and draw. It is no wonder that art appealed to us Bonner children in many forms. We emulated Mother's creative ways, and my brothers Chuck and Dana used their talent when working in the Hays museum.

Like Orv, Chuck learned collecting and preparing skills at Dad's side, but instead of pursuing a degree in geology, Chuck majored in painting. While at FHSU, he worked at the museum for five years, preparing fossils and painting renderings of prehistoric life. Chuck remembers, "Dad was pretty encouraging. Even though he didn't quite understand the idea of being an artist full-time, he liked the paintings and other artwork I made when I was at Hays." Chuck received his master's degree in painting in 1974. As the years went by, he blended art and paleontology, especially after he and his wife, Barbara Shelton, opened the Keystone Gallery and Fossil Museum in rural Logan County, across Highway 83 from Monument Rocks.

Two of my other siblings, Chris and Dana, became full-time artists. Chris earned her bachelor of fine arts from FHSU in 1984 while raising two daughters. Although she did not have the opportunity to work on museum exhibits, she did hold a job in the anthropology department at the University of Kansas before coming to Hays to get her degree. An assiduous collector of small fossils, her eye for detail served her well; she spent many years in the workforce as a graphic artist. Dana graduated from high school in 1971 and also attended Fort Hays State. He earned his bachelor's degree in drawing with a minor in painting. Then he obtained his master of fine arts in painting with a minor in photography in 1979. From sixth grade on, he knew he wanted to be an artist, and the degrees from FHSU helped him hone his craft with critical instruction. The degrees also provided credentials for professional jobs.

Chuck and Dana both worked at the FHSU natural history museum, where they combined art and paleontology. At the time of Chuck's arrival for summer school in 1968, George Sternberg was already in a nursing home in Hays, so Chuck was not able to learn alongside him the way Orv had. During that same era, J. D. Stewart did research at the FHSU museum as well, working with Myrl Walker identifying specimens. He met both Dana and Chuck there. Dana's work overlapped with J. D.'s for a full academic year. After Walker retired, Dr. Richard Zakrzewski (1940–2024) became the museum director and chief curator.[11] Dana recalls, "Chuck and I were the only work-study students outside the geology department who were working in the museum at that time. After Myrl Walker retired, I enjoyed working with Zakrzewski. He was a great teacher and encouraged artistic expression."

With Chuck and Dana in college, I was the only child left at home with Dad in Leoti. On a balmy spring day in 1972, he decided to go fossil hunting. On most of his fossil hunting days, he and Dolli set out as soon as I left for school. Now that Spiker was giving him trouble, he was using a different vehicle for his "fossil wagon," a heavy pink-and-white 1956 Pontiac station wagon. The model was a Pontiac Safari, so he said of it, "Safari, so-goody."

When I came home from school that day, he wasn't there. Normally, he would have returned home by the end of my track practice, so as it got later, I began to worry. Around six, he came to the front door with a tall, leathery Logan County rancher, who was dressed in blue jeans and a worn cowboy hat. Opening the door, Dad reeled back and forth in the doorway like a ripe head of wheat in the wind.

"Dad! Are you okay? What's going on?" I cried. "I rolled the Pontiac!" he slurred. He had taken a downhill curve of a newly sanded dirt road too fast, and he hit a ridge of loose sand. The Safari station wagon flipped, landing upside down in the pasture at the curve in the road west of Cedar Canyon. The rancher who came upon him was an old friend who had known Dad since the Whoopee Maker days. He assessed the damages: Dad had lots of lacerations and bruises that were turning purple already, and he could possibly have some broken bones or sprains. Seeing how much agony Dad was in, his rescuer gave him the only remedy he had on hand—whiskey. He wanted to dull Dad's pain so he could get him and Dolli back to Leoti.

There was also unseen pain. When the Pontiac took flight, Dad was not seat-belted in, so he instinctively grabbed on to the steering wheel. Doing so, he bruised his chest and shoulders severely on the rigid plastic wheel. At the same time, he clenched his teeth so tightly that some of the front ones cracked in half lengthwise when the car landed.

When Dad showed up at the door, I didn't know what to do. He was not making sense and he looked bad. I knew he never drank on trips to the fossil beds, so his wobbly drunken condition concerned me. The rancher politely left, leaving the situation in my fourteen-year-old hands. "You need to go to the hospital," I said to my father. "No, I can't do that. I'll just go rest in here and see how I feel in the morning," he drawled, staggering toward his room. He immediately fell asleep and began snoring loudly. I couldn't rouse him.

Frantic and confused, I ran to ask a couple of neighbors for advice. They asked if he had insurance. Of course he didn't! After talking to some adults, I decided there wasn't much I could do. My father was a stubborn man, and if he insisted on not going to the hospital, then by God, he wasn't going. I resigned myself to the situation and went home. In a panic, I remembered the dog and gave her the once-over. Dolli had a small cut under her eye, probably from a flying fossil tool, but other than that seemed unharmed.

Dad spent the next month on the couch, watching TV. His face, arms, and chest turned purple, blue, then eventually green and yellow. After a few days, we assumed he didn't have internal injuries; he would've died had that been the case. He joked about that even though he was barely able to move. Reflecting on this episode as adults, my siblings and I agreed the issue was not stubbornness but money. Lacking insurance, he couldn't afford to go to the hospital. Eventually, he healed up on his own, like an old pit bull that had crawled under the porch.

Nor could he go to the dentist. Over the next few months, he extracted the damaged teeth, one by one. He drank whiskey to numb the pain, pulled them with his strong fingers (or pliers), then applied more whiskey on the gums. He did this procedure, one broken tooth at a time. Eventually, after the last tooth was gone, he would get dentures, but he never liked wearing them except if going out for special occasions, such as to Aunt Veda's eightieth birthday celebration in Wichita in 1977. The dentures were always an issue with him. He had quoted a corny line his whole life—"Be true to your teeth, or they will be false to you"—and the adage fit, but the dentures did not. He rarely wore them, choosing to eat soft food, which he called "gumming it." With his dentures out, he could mash up his face into a comic caricature that he called Andy Gump.

Pop eventually got back to his active self. He was glad to do things besides watch television, although he did enjoy the limited shows we could bring in on NBC. His favorites were *Sanford and Son* (he identified with Fred Sanford) and *The Tonight Show with Johnny Carson* (calling Johnny a fellow Nebraskan). But now he was finally able to putter around the house, cook our meals, and work on fossils in the garage.

One particularly blustery day I heard him outside, yelling at the top of his lungs. His words were muffled by the strong wind, which was gusting to thirty miles per hour. I couldn't hear what he was saying until I opened the door and saw him, gesticulating and crying to the heavens like King Lear: "Blow, you son of a bitch! Blow my ass off!" The reason for his frustration was a large branch of the box elder tree beside the garage had broken off and was draped over the power line to the house. I asked if he was okay, and he replied, "Yeah, I'm just mad." Later, inside the house, I said, "I've never heard you do that." He replied, "I got that 'blow my ass off' phrase from my mother. She used to yell at the wind all the time."

Flashing infrequently worn dentures, Dad celebrated Aunt Veda's eightieth birthday with her in Wichita, 1977.

Even though it sometimes drove him to distraction, Dad venerated the power of the wind. He would have been happy to see the proliferation of wind farms in southwestern Kansas in the modern era. As for me, after being away from western Kansas for decades, I'm now surprised by heavy winds when they occur at my non-Kansas home. I forget that we lived with incessant wind on the High Plains, ripping down from the Rockies.

I had another harrowing incident in the house during my high school years. Like many residents of Leoti, Dad had a propane gas tank in the back yard and used that fuel for heating, hot water, and cooking. We did not have central heat, so Dad operated our propane heater manually. Its large fan blew warm air through the house.

Dad got up early on frigid mornings, went to the utility room, and cranked up the heater that blew into the rest of the house. He kept it running until the rooms got up to about eighty degrees. The huge gas burner made a big *whoosh* when he lit it, then it clicked a little while it was warming up. When he turned the fan on, we would breathe a big sigh of relief.

One time Dad needed to be out of town for the day, taking care of business at the Social Security office in Dodge City. I found out later that Dad claimed his Social Security benefit at the earliest possible time because he was desperate for money. He was forty-six when I was born and sixty-four when I graduated from high school. During my sophomore year, he filed at age sixty-two, but the monthly check was minuscule since he had never contributed payroll taxes. It was two hundred dollars a month, and he got an additional fifty dollars for his last dependent. But that amount was enough to put food on the table and buy gas for fossil hunting. Social Security was the first steady monthly income he received since renting the theater building.

Before he left, he told me, "Don't touch that heater while I'm gone." This wasn't a problem while I was at school that Friday, but toward evening, I started getting very cold, and there was no sign of Dad. So, I climbed on a chair and peered into the front side of the heater, which was installed at the top of the wall facing into the living room. I wanted to see the machine's layout and chortled when I looked through the vents and saw a couple of Old Grand-Dad empties placed along the inside door jamb. He was keeping them there until his next trip to the city dump.

I went into the utility room and stood on a chair again to see what was going on. I flipped a lever that I thought was the gas switch, then lit a match and moved it slowly under the big burner. Nothing happened for a moment, so I moved it closer, leaning in to see if the match was near the row of burner holes. The flames from the heater shot out in a larger *whoosh* than I had ever heard before. I whipped my head away from the flame and immediately smelled burnt hair. But, relieved that the burner was lit and that I was in one piece, I pulled the chain to start the fan. I let the heater run for a good hour before turning it off and going to bed. Before retiring, I took great care to trim the singed bits of hair from my hairline, eyebrows, and eyelashes.

The next day, a Saturday, Dad made pancakes and coffee with milk for breakfast, as he did so many mornings. He leaned in toward me, sniffed a little, and said, "You turned on that heater, didn't you?" I looked at him as sweetly as possible and said, "But Poppa, it was really cold in here." He frowned at me, then replied, "Honey, I didn't want you to blow yourself up, but I guess you survived. Just don't touch it—ever again!"

I am sure that he was tremendously upset that I had disobeyed and had put myself and the house at risk, but that feeling was overshadowed by relief. Later, he teased me about it. "It went *blooie*, didn't it?" he said, laughing. Looking back on this episode, I thought this was a perfect example of his hands-off parenting style, but it also showed his generous side. He understood I simply wanted to be warm.

As a parent, he didn't yell at us. He rarely even said "No!" emphatically. Instead, he'd quote things: "Nay, nay Pauline" (a phrase he had heard at an auction) meant we should choose a different option. Other alternatives to no were "I'd be afraid of it," and "We'll see." If we were taking too long to do something, the line was "Time is of the essence." Instead of saying, "Go to bed," he would say, "Bed repairs." It took me awhile to figure that one out. How does a person repair a bed? In Shakespearean English, though, repairing to do something, as in "repair thyself to bed," means the same thing as "go to bed." Dad was full of these odd phrases.

And silly songs. If I was looking a little unkempt before school, instead of issuing a direct order, he sang a children's song: "She combs her hair but once a year / Nickerty, nackerty, now, now, now / And every time it brings a tear."[12] Needless to say, I went to the bathroom promptly to do something about my hair.

During the rest of my days in Leoti, Dad stayed close to the house. Manually operated devices like the heater tied him there (that and taking care of the dog). He had a keen sense of responsibility, but he was more homebound than most parents due to the privation of our life at that time.

Privation worked out well for me when I migrated to college. Our poverty level qualified me for grants to attend the University of Kansas in Lawrence in 1975. I didn't want to go anywhere else. The summers I had spent in that delightful college city with my siblings, nieces, and nephews gave me exposure to the campus and, most importantly, to the museum.

Dad was the biggest part of my support system, but he had no hands-on experience with college. Thankfully, the summer before I took off for my Jayhawk journey, my brother Steve showed me the arcane way the class catalog (a half-inch-thick book) worked, and how to interpret its cryptic symbols. I drew a grid on an index card to visualize my Monday, Wednesday, Friday classes beside the Tuesday, Thursday ones. Then Chris gave me a tutorial on handling finances, showing me how to balance my new checkbook. It was sheer luck that I checked all the boxes for financial aid on my KU application, including the "Scholarship Hall" box. I was accepted into a very affordable housing option, Miller Hall.

As much help as I had from my family, I still made mistakes constantly. I showed up at KU's Allen Fieldhouse to "pull cards" for my classes (at that time, students walked around the arena and literally pulled a computer card from a box for each course section they wanted). Tables with card boxes were set up, but no one was in the fieldhouse but me and a lost-looking skinny boy. I walked up and asked if he was trying to enroll, and the two of us figured out that enrollment was the next day. What he had in common with me, other than being a lanky brown-haired kid, was he too was from far western Kansas. He hailed from Ulysses, in the middle of Grant County, two counties south of Leoti. It was like we had both fallen off a wheat truck.

I was fortunate to work in the Dyche Museum my freshman year of college, in the mammalogy and paleontology departments. One of my paleo projects was to catalog the bones KU scientists had found in the Natural Trap, a Pleistocene site in Wyoming. I painstakingly wrote numbers on the bones and identified them on four- by six-inch specimen cards. Then I made punch-card specimen data to move the rudimentary information onto a computer mainframe. My job at the museum was a bridge between manual recordkeeping and computerized data. These days, museums catalogue their specimens electronically; indeed, datasets provided by museums have been crucial for giving me information about Bonner fossils. A list of Museums with Bonner Fossils, as best I can determine as of this writing, is at the back of this book, page 220.

I got additional mileage from my Natural Trap work by writing my biology lab paper on the animals that had fallen to their deaths through a hole at the top of a deep cave. Happily, my lab instructor was a paleontology graduate student, and my paper got an A. In addition, I met J. D. Stewart, who was working on his doctorate. He was researching the collections and going on fossil hunts with Orv and Larry Martin. Like George and Shelly before him, J. D. knew most of the Bonner family.

Back in 1927, when Dad collected his first major fossil, the *Xiphactinus* skull, he etched his name in a chalk bluff overlooking the quarry. This group of canyons was known as the Valley of the Mosasaurs. The name "M. C. Bonner," carved under an overhang, is still there. Recording his name in chalk near the site of an important early fossil is another thing Dad shared with George Sternberg, who etched his initials near his first plesiosaur in 1892.

In the vastness of the wilderness, it is a combination of humility and pride that makes human beings—the latecomers on the scene—want to make their mark, a reminder of the present day among prehistoric bluffs. "I was here," the effort declares.

KU's Samuel Williston noticed that human etchings withstood the test of time in the chalk. In 1897, he wrote, "My first knowledge of the Rock [Monument Rocks] dates from October 1874, and since that time I have seen but little evidence of erosion. In various places throughout the chalk beds of the Smoky Hill River I have observed marks scratched by myself eighteen years previously that appeared as clear almost as when they were made. The erosion in general is not nearly so rapid as one would think."[13]

After he left Leoti and settled in Healy, Dad was looking toward new horizons. Where he had once been a fixture in Wichita County, with his recognizable fossil wagon heading north on Highway 25, he now was becoming well known in Lane County. The eccentric old gentleman who worked on fossils in his home would be interesting fodder for regional newspapers the rest of his life. One example of this local coverage was a photo-essay called "A Day in the Life of Shields," published in the *Dighton Herald* on June 21, 1989. Along with pictures of the Shields postmaster, a man filling a propane tank, and an elderly couple at home, Dad is shown working on a fossil in his Quonset hut. Just a normal day.

In 2017, Chuck sat in front of Dad's name, carved below a chalk ledge in 1926 above his first Xiphactinus skull quarry. Photo by Kris Super.

Two other family-loved stories we have kept through the years were "Full Time Fossil Finder," by Kathy Hanks, a spread published in *Kansas!* magazine in 1983, and "Fossils Feather His Cap," by Sharon Hamric, a feature that appeared in the *Wichita Eagle-Beacon* in 1985.

Mystery Fish
1970s–80s

In March of 1971, I was an eighth-grader, Dana was a senior, and Chuck was home from Fort Hays State University for spring break. That week we also had visitors from the University of Kansas, Orv and his good friend Bob Patterson, who worked as a zoologist in Dyche Museum from 1957 to 1992. Bob was another story-telling character. He was red-haired and red-bearded, with the ruddiest complexion I had ever seen. His face grew animated when he described adventures with animals. I remember vividly his tale of capturing a snapping, snarling coyote. Also a big kidder like Dad, he dubbed me "Mel-ornery."

During the week, while I spent time at a slumber party with friends, the fellas—Dad, Dana, Chuck, Orv, and Bob—went to the Big Place. Bob was photographing wildlife and looking for living species such as prairie dogs and pack rats (also called gray woodrats in western Kansas). Pop enjoyed capitalizing on pleasant weather to do as much hunting as possible, taking advantage of all the extra sets of eyes to locate fossils. He would return later to dig on the specimens to determine if they were worth collecting.

Chuck and Dana spread out in opposite directions to hunt the high bluffs. Right away Dana found a beautiful, almost complete *Ctenochelys* turtle on a towering bluff of orange-ish chalk. This turtle, named a separate genus by Charles H. Sternberg in 1904, is much smaller than *Protostega* or *Toxochelys*, measuring only eight to ten inches long. What made the collection of this turtle memorable, Dana recalls, was Dad held on to Orv's belt as he ventured out on a foot-wide ledge to examine the fossil. Orv stayed out on the ledge to collect it but first chopped himself a better foothold in the hard bluff. This would not be a fossil Dad would come back to collect: It took a Mountain Goat collector to bring in the discovery of a Mountain Goat hunter. Orv has an eight-by-ten photo Bob Patterson took of him checking out the turtle high up on the side of the sheer, peach-colored bluff, with Dad standing, a bit nervously, on surer ground behind him.

This day was propitious, not only for the high perch from which Orv collected the turtle but also for the discovery of what the Bonners would call the mystery fish. Chuck recalls, "I was really jealous that Dana found such a great turtle, and all I found was what I thought was a spread-out deposit of *Protosphyraena*, a fish that is abundant in the chalk." After Orv finished collecting Dana's turtle, he and Dad carefully made their way out to the high isolated bluff where Chuck had found the large fish. Dad began digging on it. "Damn, these bones look weird. Never seen a *Protosphyraena* this big!" he commented while working.

The top of a steep knoll was a difficult area to work in, and they knew that hauling down the fossil in heavy slabs or casts from that height would be tricky. The scattered bones were so unusual that Dad decided to go home and do a little research, then return later to get them. He covered the fossil with pieces of chalk and marked the site with a "sheepherder's cairn" so that he could find it later after his sons had gone back to their homes in Hays and Lawrence. Returning to the quarry by himself to dig out the specimen was no problem, but based on its precarious location on the high outcropping of yellow chalk capped by lichen-blackened rock, he knew he needed extra manpower to collect the fish later that summer.

I spent most summers from junior high on with my brothers' and sisters' families. Dad believed I would not have much to do in Leoti during the summer and that being exposed to (and nurtured in) different environments would help my overall development. Plus, he got a few months' break from me! The summer of 1971, which was right before my freshman year of high school, I had spent most of June in South Dakota with Clare Jane's family, then most of July in Lawrence at Orv's house. Both Clare Jane and Orv had four children, so there were plenty of nieces and nephews to play with.

When I was away from Leoti during the summer, Pop and I were avid pen pals, me addressing the letters to "Little Poppa." In 1971, I received a birthday card with a missive from Dad. In addition to chatty pleasantries, he wrote about the mystery fish, which he had decided was not a *Protosphyraena*. He was now calling it a "skate." By the time of the letter, he had collected the skull material but needed to go back for the rest of it. Parts of his note read:

> Dear Mel: Happy Birthday!! How are you? Hope fine. I'm OK and still working on the skate. Went and visited Dana & Chuck & took the skull down for Walker to see; he doesn't know what it is either. Got the rest of the specimen in a plaster slab last Monday and got it rained on—don't think it hurt it tho. Have started digging out

from under it now but get rained out every day. ... Well, I'd better go. This letter will do my arm up & I may not be able to work on the fossil. Ha. ... All my love to you, Dad.

Not long after that, with the help of his Leoti friend Charlie Norton (1942–2024), Pop extracted the rest of the large fish. The collection of this fossil was epic because of its size and its almost inaccessible position. It was so heavy that during the first attempt at collection, Spiker broke one of its steel springs. It was the "end of an era" for the fossil wagon. So Charlie used his pickup truck to bring the specimen from the quarry to Dad's garage in Leoti.

Charlie recalled the challenge in an email he sent me in 2012:

> The fossil that I remember the best was the large skate or ray that one of the boys had found on the east side of the Big Place. We took a bunch of old tires to protect the fossil casting as we let it off the high ledge. I had several lariat ropes on it, and Skeet always said, "We brought it out on a Diamond Hitch." We chopped out a path through the choke cherries and leveled a smoother trail for Ol' Spiker, but even at that we broke a spring from all the weight and roughness. I came close to getting my leg broke that day while bringing out the largest plaster casting.

Charlie Norton was a western artist from Leoti, well known for his bronze sculptures. The monumental Buffalo Bill statue in Oakley is a famous example of his work, and he created a bronze bust of his dear friend, my Pop, which he called "Skeet," in 1989 (see photo on page 5). Charlie, who graduated from high school the same year as my brother Steve, was always friendly with Dad but male-bonded with him more after Chuck and Dana left Leoti. In the same email describing the collection of the "large skate or ray," Charlie remembered Dad nostalgically:

> I liked to go to the fossil beds with him and on one occasion he started [reciting] poetry ... "What are we all but the potter's clay / To wither and die and fade away. / Some must lose and some must win, / But aren't we all equal under the skin?" One thing for sure, Skeet left a legacy that will not fade away. ... I worried about Skeet working out in the hot summer heat and being by himself, but he was a tough "Ol' Grizz" and never gave up. He was like a father to

us—we will always love the guy and get emotional when thinking about all the great times with him.

Once collected, the large fish stayed in Dad's garage in Leoti for more than three years. The collected specimen included pectoral fin material and multiple bones from the skull. He continued to chip away at it, always wondering, "What the hell is it?"

Bones of the Bonnerichthys gladius type specimen in situ before collection in 1971. The genus was named by scientists in 2010. Photo by Charlie Norton.

Consulting with Orv, Dad decided he wanted the singular fossil to go to the KU department of vertebrate paleontology where it could be researched fully. There it was originally identified as *Protosphyraena gladius*. Dr. Larry Martin was in charge of acquiring the fossil in 1975 and loaning it to scholars who wanted to take a closer look at it.

In 1990, Dad attended his last Society of Vertebrate Paleontology meeting, which was held in Lawrence. Chuck drove Dad to the meeting from western Kansas, and while neither of them was a dues-paying member of the society at that time, they were recognized by many and welcomed to the event. A third Bonner, Orville, helped coordinate the meeting at the University of Kansas. During the program, Dad met some of the new paleontologists at the museums that held old Bonner fossils.

Pop saw Shelly Applegate at this meeting and told him about the bizarre fish in the KU collections. At one break in the sessions, Shelly and the Bonners went down into the basement paleontology lab where Orv worked. Dad had been telling the fish expert about the unusual fish that he collected in 1971.

They all examined the bones of the "weird fish," lying in the casts it had been collected in. Shelly perused it a bit and then said he had no idea what the fossil was. It wasn't a shark, a skate, or a *Protosphyraena*. "I'm serious, this has to be new to science," Dad told his friend, his eyes crinkling. "That crazy thing doesn't look like anything I've ever seen before. It's so huge. Look! That sclerotic ring [bones around the eye socket] is almost as big as a volleyball!"

Twenty years later, the identity of the unidentified fish would be explained, but neither Shelly nor Dad would be around to see it. Dad died in 1992 at the age of 81, and Shelly died in 2005 at the age of 76. The last time Dad saw Shelly was at that SVP meeting in Lawrence. By that time, Shelly had moved to his final position in Mexico City.

A decade later, another major fossil fish discovery occurred a few weeks after Dad's seventy-first birthday. This was a different kind of fish, one well known to Dad, a *Xiphactinus*. But this specimen held a surprise.

On a sunny June day in 1982, Dad, Chuck, and Dana, accompanied by family and friends, decided to head out to some canyons in southeast Logan County. Chuck and Barb's son, Dad's youngest grandson, was a baby in diapers. He too was growing up in the fossil beds.

After all the hunters got their tools, they spread out in different directions. Dana noticed a couple of pack-rat nests near the bottom of a gully as he moved toward a desirable knoll. Pack rats are rarely seen during the day but leave evidence of their nocturnal activity: heaps of gray sticks and black dung pellets littered around a hole dug into the soft earth. Dana jumped over the nests and continued up the gully.

Dana has a meticulous way of doing everything, and this includes fossil hunting. He paced himself as he approached what looked like a promising spot. He walked slowly around the curves of knolls, scouring around the top layers, then the middle ones, then the lower ones, moving his eyes over the sections methodically. In the vertical middle of the bluff, he noticed about six inches of exposed fish tail. The lobe of the tail fin was large, so he assumed it was a *Xiphactinus*. Dana positioned himself on the knoll above the fossil, then started excavating above the tail. He dug for half an hour, determining the tail was angling back into the wall toward the rest of the body of the immense fish.

"It was fairly hard chalk," he remembered. "I dug on it for a while—it was that hard yellow stuff that takes quite a bit of effort to remove with the pick. After a time, I heard Dad coming up the gully, and he asked what I was digging on up there. I told him I thought it was a *Xiphactinus* tail, and he made his way up. After that, Dad took over, as he always did," Dana said, laughing.

Pop scraped and picked away for most of the afternoon, taking small breaks for drinks of water. He decided to excavate the quarry back into the wall narrower than normal due to the hardness of the chalk and the depth of the fossil inside the bluff. Also, Dad was now in his seventies, and digging all day took much more out of him than it had in the early days. With Chuck and Dana's help excavating the hard chalk, the quarry ended up being seven feet deep. "If this is all here," Dad thought, "It's going to be hell hauling this sucker back to the car." The bluff was situated a good three hundred to four hundred yards away from the closest place a vehicle could park on the prairie above.

A hairy hippie in 1982, Dad works in the narrow fish-within-a-fish quarry as Chuck, Dana, and baby Logan look on. Photo by Barbara Shelton.

Dad decided to take the specimen in two pieces. The heavy work was removing the overburden, but another part of excavation required careful scraping to take off as much chalk as possible to make it lighter and to barely expose the bones. It was important to locate all the bones but not delve too deeply. While he was clearing off this shallower layer, Dad noticed bones of a different fish inside the *Xiphactinus'* belly. The bones were of *Gillicus*. "I'll be damned!" he thought. "Could this be another fish-within-a-fish like George's?"

Pop and my brothers used both proven methods of collection for this specimen. They encased the tail section in a burlap-and-plaster jacket and collected the rest in a large framed slab. Dad knew that even with Dana and Chuck helping, he needed to get some extra muscle to bring this specimen from its quarry. For this *Xiphactinus,* his volunteers were two young friends, football players from Healy. The haulers owned a heavy-duty pickup truck and were more than happy to help the Bonner brothers grapple the large plaster jackets into their truck bed. The boys removed the slabs one at a time, taking them to Dad's place in Shields and placing them on his worktable in the Quonset hut.

After Dana and Chuck returned to their homes, Dad worked on the specimen in the lab and confirmed that the fish inside it was complete and not just some scattered parts. Next, he got on the phone with connections who could find a good buyer for the fossil. As usual, he wanted the stellar fossil to go to a prestigious museum. At the time, this was the second known complete fish-within-a-fish from the Niobrara chalk. At nearly fifteen feet long, wider, and slightly curved, the Bonner specimen was larger than the famous Sternberg fish-within-a-fish, which is straight and measures fourteen feet. The complete *Gillicus* inside the *Xiphactinus* is upside down and slightly curved also, whereas it is straight and right side up in the Sternberg specimen.

Since the 1980s, more fish-within-a-fish fossils have been found or located in museum storage. Because there are a handful in the fossil record now, this confirms one of the ways the ferocious fish met its death. It's easy to imagine the massive predator, its internal organs punctured by the smaller fish's thrashing fins, dropping to the ocean floor with the fish that took its life, landing heavily, and rapidly covered by silt.

Dad eventually sold the fish-within-a-fish to the Royal Tyrrell Museum in Alberta, Canada. At that time, the newly built facility was acquiring all kinds of items, historic and prehistoric. Today the Canadian museum displays the Bonner fish-within-a-fish in its Dinosaur Hall—a fitting home for the Kansas fossil because *Xiphactinus* was swimming the Cretaceous Sea at the time large dinosaurs like *Albertosaurus*, named after the museum's home province of Alberta, were roaming the land. The institution is nestled within the Late Cretaceous fossil beds of Alberta's Horseshoe Canyon Formation.[1]

In 1986, Dad met a crew of videographers from Fort Hays State University who wanted to film a short documentary about him. The year he turned seventy-five, KOOD, the public television station based in Hays, now under the Smoky Hills PBS affiliate, aired a program about his time in western Kansas and his experiences as a fossil hunter. After interviewing Dad extensively, the show's producer and director, John Stoss, named the half-hour program *Marion Bonner: Just Another Vertebrate.*[2] The title derived from the last scene of the show, where Dad is lying down, as if dead, on the chalky ground near Monument Rocks and eerie music is playing. Then Dad suddenly wakes up, throws his head back, toothlessly cackles at the camera, and proclaims, "Aww—You all worry too much! After all, we're all just a bunch of vertebrates."

In a middle section of the program, Dad walks around the Fort Hays Natural History Museum on the college campus, wearing his trademark flat felt newsboy hat, his white hair long in the back. He points out Niobrara Cretaceous fossils with his middle and missing index finger. He talks about his donation of the short-necked plesiosaur to the museum and recalls how "My late wife, Margaret, said I disowned the family over that fossil." His eyes sparkle, and he chuckles, remembering her comment. He also points out "George's famous fish-with-in-a-fish."

The show captured Dad's fondness for local history and prehistory but also his sense of humor. Some parts are downright silly, but its airing brought him a little local notoriety. This exposure helped contribute to the "Old Man of the Fossil Beds" legend, which was a double-edged sword. He enjoyed the recognition, but it made him more defensive and protective of the fossil beds. Unfortunately, this program is not viewable online.

Just Another Vertebrate was an appropriate title that summarized Dad's philosophy. He viewed the constructions of human civilization, such as the development of religion, literature, and the arts, as valuable but sometimes destructive attainments. He always professed, "Man is first and foremost an animal. It's when we forget we're part of the natural order of things that we get in trouble." He continually repeated the notion that civilization is but a thin veneer over Earth's long history.

The strange fish that had lain unidentified for years in the KU vertebrate paleontology collection was finally described by paleontologists in 2010. It would become a new genus: *Bonnerichthys*.

While a doctoral student at the University of Kansas, paleontologist Anthony Maltese took great interest in the fish Dad called "weird." After Maltese relocated to the Rocky Mountain Resource Center, he asked Dr. Larry Martin if he could prepare it further and make a cast of it. At about the same time, a visiting graduate student from Oxford University, Matthew Friedman, saw a resemblance between the Kansas fossil and other filter feeding giant fish he had studied in Europe. In 2010, a group of paleontologists joined with Friedman to expand the research, publishing it in *Science* magazine and noting that "large-bodied suspension feeders (planktivores), which include the most massive animals to have ever lived," were absent from the marine environments of the Cretaceous. This niche, where animals grow gigantic on a diet of plankton, is occupied by whales in the modern era.[3] They named the type specimen *Bonnerichthys gladius*.

Two of the article's co-authors, Kenshu Shimada and Michael Everhart of the Sternberg Museum, provided an additional nearly complete specimen for study. Everhart explained their involvement with the scholarship that described the Cretaceous *Bonnerichthys* ("Bonner fish"). In *Oceans of Kansas,* he discussed the material found by Dr. Shimada and that he himself collected with the assistance of other scientists for the Sternberg Museum:

The remains were then sufficiently prepped out to be included in our paper in *Science* describing *Bonnerichthys gladius,* the new genus of filter-feeding fishes from the Western Interior Sea (Friedman et al. 2010). Two years later we did a follow-up report on the distribution of *B. gladius* remains across North America (Friedman et al., 2013).[4] There's still more work to be done on this previously unknown fish from the late Cretaceous, including a description of the pectoral fin, the teeth (yes, tiny ones) and the identification of the stomach contents that were discovered by Anthony Maltese during preparation.[5]

Now that *Bonnerichthys* occupies its own niche and genus, museums around the world are reexamining fish material previously attributed to other genuses, sometimes *Xiphactinus* but most often *Protosphyraena*. For example, a huge, perfect pectoral fin collected by George Sternberg, on display in the Sternberg Museum and originally labeled *Protosphyraena,* is now identified as *Bonnerichthys gladius*. Scientists have further determined that the first remains of *Bonnerichthys*, called *Protosphyraena*, were originally discovered by Professor Benjamin F. Mudge of Kansas State University and the Kansas Geological Survey.[6]

The article by Friedman et al. in *Science* magazine was a major coup. Closer to home, newspapers covered the story. "Fossil 'Discovery' Rewrites History," was the headline of one piece in the *Hays Daily News* on February 21, 2010, that quoted both Chuck Bonner and Mike Everhart. Summarizing the information for a lay audience, Mike Corn wrote, "The late Marion Bonner was right: The discovery in 1971 by his son Chuck, then 21, was indeed unusual. ... On Thursday, scientists announced that it was deserving of its own genus, proving to be something of a missing link between the oceans of 100 million years ago and today." Later in the article, Everhart noted, "It is the biggest bony fish feeding in the Cretaceous sea." The discovery that it was a huge fish that fed on plankton "filled in the blanks."[7]

We Bonners talk amongst ourselves about how tickled Dad would have been, had he lived to see this fish described as filling a missing niche. The idea of a new species excited him, so a new genus would have been exhilarating. The appellation *Bonnerichthys* was given to honor the Bonner family—Chuck for finding it, Dad for collecting and preparing it, then pushing for KU to have it, and Orv for getting it there. Between ourselves, we also think it's funny that our name is attached to the "big-mouthed filter feeder" of the Cretaceous. Something feels so right about that.

The thirty-nine-year timeline of *Bonnerichthys* went like this: Discovered in 1971, collected that summer, taken to the KU natural history museum in 1975, examined and puzzled over for years, rediscovered and described as a new genus by scientists in 2010. (The cover photo of this book, taken by Charlie Norton, shows Dad at work in the 1971 *Bonnerichthys* quarry.)

In the 1970s and 1980s, after moving to Healy, Pop continued to expand his association with other museums and fossil buyers. In 1979, he made his mark again, with the discovery of his last major fossil, a *Tylosaurus*. He laid out the complete specimen of the *Tylosaurus* on his long kitchen table, adjoining buffet, and library table.

Complete specimen of Tylosaurus prorigor, twenty-seven feet long, discovered and collected by Marion Bonner in 1979. Photo courtesy of Memphis Museum of Science and History.

Dad sold his "outstanding, nearly complete Tylosaur" to a fossil agent who assured him it would be placed in a museum. The fossil eventually ended up in Memphis, Tennessee. Rose Basom, the former curator of natural history at the Museum of Science and History in Memphis, confirmed in an email to me in February 2023 that the *Tylosaurus* was purchased from a broker in 1979 and was "discovered by Marion Bonner of Healy, Kansas." The museum website's description of the fossil states:

> Our display specimen came from the chalk beds of Kansas, but mosasaur fragments have been found in Tennessee, Alabama, Mississippi, and elsewhere in the Mid-South. Our curators have excavated specimens including *Prognathodon overtoni*, *Plioplatecarpus depressus*, and *Mosasaurus maximus*. Some mosasaur species grew to be longer than 50 feet. Our specimen, a *Tylosaurus*, measures 27 feet, 6 inches. It is unusual because it contains 95% of the

original bone structure. As a predator, [the mosasaur was] the aquatic equivalent of the T-Rex. Like the dinosaurs and other giant reptiles, mosasaurs became extinct at the end of the Cretaceous Period in a world-wide extinction event.

After this outstanding fossil, Pop found some partial specimens but mainly served as middleman to museums by preparing and placing fossils found by Dana and Chuck. Toward the end of his life, he formed a full partnership with Chuck that was reminiscent of the relationship between Charles H. Sternberg and his son George.

By August 1989, Dad paid off the remainder of his small mortgage on the land, house, and Quonset hut in Shields. He originally bought the Shields property in 1981, writing out the contract himself in his emphatic longhand. He told me in a phone conversation that the lawyer and banker looked at his hand-written contract and said there was nothing wrong with it. All they added was a description of the property. "Hell, I had commercial law in high school," he said. "The only thing they asked for the day of the sale was for me to insure it. I said, 'No, I won't buy it if that's the case.' " Even though the Lane County banker and lawyer thought that an uninsured property was highly irregular, they relented.

Just as he had no concern about retirement savings (the concept would have baffled him), Dad eschewed insurance his whole life; always saying it was a "racket." He never insured any of his houses, nor did he have any health insurance until he got on Medicare. Who needs insurance when you live close to the bone, viewing life through an ironic, somewhat fatalistic lens? And who needs insurance when you're living by your wits, persevering and surviving any casualties that might occur? And you're proud of it.

The Fossil Record
1980s–90s

FOR ALL THE YEARS Dad spent combing the Niobrara "graveyard," he saw so many scraps of *Inoceramus* shells, pieces of *Ostrea*, and other seashell fragments that he tended to ignore these invertebrates in favor of the vertebrates. Yet the two species named for him (one of them now a junior synonym) were of invertebrates. In 1968 he found the first *Pecten bonneri*, which had a shape like a thick modern scallop but lacked the scallop's long, radial ridges. Maybe he noticed it because it was a whole seashell—a complete bivalve mollusk. He took it to Myrl Walker in Hays to see if it was "new to science."

Walker showed it to Fort Hays State University geology professor Halsey Miller, who had described the two squids *Niobrarateuthis bonneri* and *Enchoteuthis melanae*. In a comprehensive paper about Kansas Cretaceous invertebrates, Miller designated the new species *Pecten bonneri*. In the description, Miller wrote that the fossil consisted of "both valves of a nearly smooth form, ornamented with fine, concentric growth lines" and that the specimen "was discovered by M. C. Bonner three miles south of Russell Springs."[1] The fossil shell was small, only 47 millimeters long, or 1.85 inches. This new species turned out to be another reward for the obsessive hunting he did after Mom's death. The bivalve *Pecten bonneri* is shown on page 159.

The other big reward after Mom's death, the complete *Platecarpus*, was a gift that kept giving. Pop had noticed that this major fossil was not mentioned in stories about the "Bonner Wall" in the Los Angeles County Museum. The fossil was not located until 1988 when J. D. Stewart found it in the museum's storage area. That discovery itself was newsworthy at the time, since J. D. was able to recognize that several sections of mosasaur material were in fact a single specimen and that the fossil had the "best preserved external features of any in the world."[2] The fossil was so well preserved that it had its scales, skin, and the contents of its last meal in its stomach. In coming decades, more revelations would come from the "denizen of the deep," as Dad called the *Platecarpus timpaniticus*.

Multiple scholars have analyzed the specimen since Stewart's rediscovery of it in 1988.[3] They have found signs of internal organs—reddish areas inside the skeleton may represent locations of the heart, lungs, and kidneys. The trachea is also preserved, along with part of what may be the retina in the eye. The anatomy of mosasaurs had been compared to monitor lizards in the past, but the L.A. *Platecarpus* shows the placement of the kidneys farther forward in the abdomen than in monitor lizards, a location that is more similar to whales. These conclusions were made by a Swedish-backed research initiative in 2010, which used ultraviolet light to examine the scale impressions and the tracheal and bronchial rings.[4] Theories have emerged about how the reptile swam, what it ate, and whether it was active in both salt water and fresh water. A summary of conclusions may be found in Wikipedia's page on the genus *Platecarpus*:

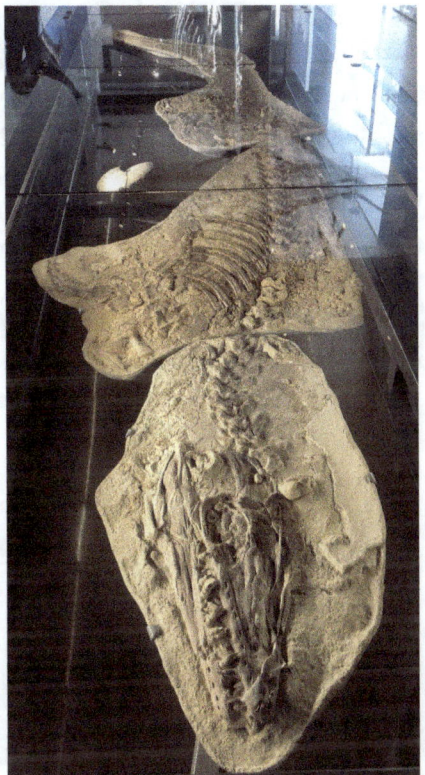

Dad's exceptional Platecarpus, collected in 1967, is on display in LACM's Dinosaur Hall. Photo by Judy Ryser.

A well-preserved specimen of *Platecarpus* shows that it fed on moderate-sized fish, and it has been hypothesized to have fed on squid and ammonites as well. Like other mosasaurs, it was initially thought to have swum in an eel-like fashion, although another study suggests that it swam more like modern sharks. An exceptionally well-preserved specimen of *P. tympaniticus* known as LACM 128319 shows skin impressions, pigments around the nostrils, bronchial tubes, and the presence of a high-profile tail fluke, showing that it and other mosasaurs did not necessarily have an eel-like swimming method, but were more powerful, fast swimmers. ... Isotopic analysis on teeth specimens has suggested that this genus and *Clidastes* may have entered freshwater occasionally, just like modern sea snakes.[5]

Pop would have been astonished at the continuing scientific advancements in analyzing not just the mosasaur but also other fossils, including the sharks. Among the other Los Angeles fossils, the *Squalicorax* that my brother Steve found in 1963 was included in a comparative analysis by shark experts in 2005. This study affirmed that the size of this genus could be placed around two meters long, not to exceed three meters. The authors also suggest this shark could swim very quickly and had a keen sense of smell but weak eyesight, based on morphological features.[6]

Another significant shark in Los Angeles, which garnered analysis in 2011, is Mom's *Archaeolamna*. Because it's the most complete skeleton of the *Archaeolamna kopingensis* species, paleontologists have been able to draw conclusions about its age, size, and eating habits. A portion of the abstract of a paper by scientists from Alberta, Chicago, Kansas, and California reads:

> The specimen includes portions of the upper and lower jaws with articulated teeth. ... The dental sequence of *A. kopingensis* is unique among both extinct and extant lamniforms [an order of sharks that also includes *Squalicorax* and *Cretoxyrhina*]. Associated with the jaws are fragments of the neurocranium and multiple vertebral centra. A sagittal section through a centrum shows that this shark deposited eighteen annual marker bands after its birth and adult size was attained by the 10th band. The robust but penetrating tooth morphology and large jaw circumference suggest that *A. kopingensis* likely fed upon large prey items.[7]

In 2018, an unusual Bonner fossil led to an analysis of the behavior of predation among sharks. In the composite *Pteranodon* skeleton on display in Los Angeles, paleontologists noticed an embedded tooth of a *Cretoxyrhina* shark. Responding to their query, J. D. Stewart said he had confirmed with Dad that the vertebra was found that way—with the tooth lodged in it. Then scientists later examined it as an example of how the large sharks fed on *Pteranodons*.[8]

Discoveries like these provide windows of speculation into the actions of extinct animals. Questions arise about how the shark bit the flying reptile. Did the *Pteranodon* fall into the water, or did the shark grab it out of the air, *Jurassic Park*–style? Or possibly, as calmer paleontologists, including Mike Everhart, suggest, the tooth may have detached from the shark's jaw as it was feeding on the *Pteranodon*.

To establish a type specimen, paleontologists must research and publish their findings in peer-reviewed scientific journals. Holotypes are not set in stone but may be overruled, as happened with *Niobrarateuthis bonneri*. Another name change occurred with the *Nyctosaurus* at Hays that Orv included in his master's thesis, which he wanted to name *Nyctosaurus sternbergi*. Others thought that specimen was a *Nyctosaurus gracilus*, and then, paradoxically, the name was changed again—to *Nyctosaurus bonneri*, by Halsey W. Miller, who named it after Orville. That name is shown on the exhibit sign.

However, that name is also challenged. Some paleontologists argue that all *Nychtosaurus* fossils are actually small *Pteranodons* and not a separate species at all. For the final word (for now) on this topic, Michael Everhart summarizes current scholarship, stating, "The taxonomy and systematics of *Pteranodon* were revised by [S. Christopher] Bennett, who noted again that '*Pteranodon sternbergi* seems to be ancestral to *P. longiceps*'... and that '*Nyctosaurus bonneri* is a junior synonym of *N. gracilus*.'"[9] Bennett, considered one of the foremost authorities on *Pteranodons*, argues that variations in size and crest types among these flying reptiles can be attributed to sexual and maturation differences, which he contends is not enough evidence to distinguish separate species.

Science continually corrects itself.

One of the shortfalls of field paleontology from the earliest days is that so many of the fossils collected from the Smoky Hill Chalk and Pierre Shale lack specific stratigraphic information. The full deposits of Smoky Hill sediments total two hundred meters, or approximately six-hundred fifty feet in thickness.[10] Moreover, the early fossil collectors used only generalized terms like the "Niobrara" to describe the strata of collected fossils. While fossil collectors like the Sternbergs, and later the Bonners, could identify the exact location of a fossil excavation site geographically (township, range, coordinates, and the like), they didn't identify where the fossil occurred in the stratigraphic column.

This changed starting in 1982, when Donald E. Hattin, professor of geology at the University of Indiana, published a detailed study of a full, measured section of Smoky Hill Chalk using bentonite (a type of clay with volcanic ash) as a marker and dividing the chalk into five zones based on the occurrence of certain inverte-brate species.[11] To understand the period of chalk deposition, Hattin calculated that one inch of chalk represents about seven hundred years of time. A foot of chalk, then, would be deposited over eight thousand and four hundred years, or roughly the number of years of human "civilization."[12]

So if a paleontologist finds two fossils on the same bluff that are in layers only six inches apart, he or she can tell a significant amount of time passed between the deaths of the two animals. Dad would have loved this. I can just hear him crowing, "Human culture equals a *foot* of chalk deposits!"

In 1988, J. D. Stewart expanded the understanding of Niobrara stratigraphy by examining vertebrate fossils.[13] His paper summarizing the recent history of strati-graphic study was included in the SVP's *Niobrara Chalk Excursion Guidebook,* edited by S. Christopher Bennett from the University of Kansas. The subtitle of the *Guidebook* reads: "Commemorating the Fiftieth Anniversary of the Society of Vertebrate Paleontology and One Hundred Years of Vertebrate Paleontology at the University of Kansas." Obviously, this dating of fossils from the chalk was long overdue, but the dating of some significant specimens is not altogether lost: Modern paleontologists have been able to locate the quarries of Cope, Sternberg, and others, which enables them to document the stratigraphy of those sites.

One of the last important Bonner fossils Dad sent to J. D. Stewart in Los Angeles resurfaced in 2011. Lying expectantly in museum storage, this one had been taking a "pregnant pause," so to speak. The fossil was a specimen of a *Polycotylus* plesiosaur, a relative of the short-necked plesiosaur *Dolichorhynchops* that Dad donated to Fort Hays State. The story of the *Polycotylus* in L.A. resumed, but with a surprising plot twist, when scientists at the Los Angeles museum decided to delve a little deeper into the specimen.

According to a newspaper article in the *Los Angeles Times* in 2011, more than twenty years had passed since Dad had shipped the specimen to Los Angeles in several separate burlap and plaster jackets. J. D. Stewart had left the museum by the time two other paleontologists, F. Robin O'Keefe, from Marshall Univer-sity in Huntington, West Virginia, and Luis Chiappe, the new director of the Dinosaur Institute at the Natural History Museum in Los Angeles, decided to take a closer look before displaying it as part of the Los Angeles museum's new Dinosaur Hall. The *Times* reported that:

The scientists noticed a constellation of small bones spilling over from the larger fossil's abdomen that appeared to be miniature versions of the adult ones. The similarities suggested that both sets of remains were from the same species. ... A flat seashell-shaped bone—part of the fetus' pelvis—rests on the inside face of the mother's shoulder bone, indicating that the baby was growing inside its mother when she died.[14]

The journey of a pregnant plesiosaur began after Chuck's discovery of it in 1987. He and Dad reduce the weight of the rock in one of the plaster jackets with Marsh picks while Logan supervises. Photo by Barbara Shelton.

The discovery that the short-necked plesiosaurs gave birth to live young was an exciting revelation. Prior to this, Cretaceous marine reptiles were assumed to be egg-producing rather than viviparous. The pregnant plesiosaur added new data and new lines of inquiry. When Dad sent the *Polycotylus* to the Los Angeles museum, he and Chuck both knew the smaller plesiosaur bones were inside the mother.[15] J. D. Stewart also knew, which is what prompted him to recommend that the museum acquire the fossil. After O'Keefe and Chiappe's paper was published, the fascinating information captured the public's attention, even though the Bonners' and J. D.'s knowledge of the bones was not mentioned in the *Times* story.

"Polly the Pregnant Plesiosaur," Chuck's cutesy name for the *Polycotylus*, marked a transition in Dad's collecting. Chuck discovered the specimen, dug it out, and collected it. Dad prepared the fossil in his last laboratory in Shields. After preparation, Dad did the "brokering," contacting Stewart at the Los Angeles museum, then crated and shipped the fossil.

Chuck's discovery of this plesiosaur was quite literally a stroke of luck. He was in the chalk bluffs near his home hitting golf balls with a friend in the pasture. As they were whacking away, one of Chuck's shots sliced into a canyon. Chuck traipsed down into the Pierre Shale draw to retrieve it and spotted the bones of the plesiosaur. Polly has become a star in the L.A. museum's Dinosaur Hall as well as in the virtual tour on the museum's website.

In a sense, the *Polycotylus* represented a passing of the torch between father and son in regard to "doing the heavy lifting," as Dad put it. My brother Chuck found most and collected all of the fossils toward the end of Dad's life. The only remaining Bonner in western Kansas, Chuck is the last of the family to carry on the tradition of fossil hunting and collecting.

The pregnant Polycotylus plesiosaur on display shows the bones of the fetus near the center behind the mother's paddle. Photo courtesy of the Museum of Natural History, Los Angeles County.

Short-necked plesiosaurs, whether *Dolichorhynchops* or *Polycotylus*, not only played a significant role in the Bonner family story but also paddled down the timeline of Kansas paleontology. William E. Webb, the author of *Buffalo Land*, the book Dad became obsessed with in the 1960s, acquired the type specimen of the first polycotylid plesiosaur, *Polycotylus latipinnis*. It was described by Edward Drinker Cope in 1869. At the University of Kansas in 1900, Samuel Wendell Williston named the type specimen of *Dolichorhynchops osborni*, discovered by the young George Sternberg, whose father, Charles H. Sternberg, sold this ple-siosaur to KU.

Then, in 1955, Dad found the nearly complete *Dolichorhynchops osborni* and donated it to George Sternberg's Fort Hays State University Museum. It was that fossil that bonded Dad with George and kick-started Orv's career in paleontology. In 1987, Chuck located the pregnant *Polycotylus* while playing golf in a pasture. And finally, *Dolichorhynchops bonneri*, named for Orville, surfaced. For this variety of plesiosaur, its proponent, Dr. Dawn A. Adams, believed a new species could be based mainly on the paddle structure and suggested that this short-necked species was the speed demon of the Creta-ceous seas. Her proffered name was *Trinacromerum bonneri*,[16] but the genus was changed by F. Robin O'Keefe in 2008 to *Dolichorhynchops bonneri*.[17] Regarding the species name, Adams wrote: "The name proposed for the new species is in honor of Orville Bonner, preparator at the University of Kansas. Mr. Bonner has devoted most of his life to the collection, preparation, and study of the rich faunas of the Niobrara Chalk and Pierre Shale. His consid-erable preparation skills and wide knowledge of Great Plains paleontology have proven invaluable to graduate students at the University of Kansas for many years."[18]

All members of the Polycotylidae family, which includes the genuses *Poly-cotylus, Edgarosaurus, Dolichorhynchops,* and *Trinacromerum,* have similar body types. The short-necked, thick-bodied reptile that looks somewhat like a sleek turtle without a shell, helped make connections between fossil hunters of many eras, and was of particular importance to the Sternbergs and the Bonners. The ancient plesiosaur links Webb, Cope, Williston, the Sternbergs, and the Bonners, and ties father and son fossil hunters together. Plesiosaurs also connect the two major vertebrate paleontology museums of Kansas. "Pollies" and "Dollies" dominate our lives.

Throughout his life, Dad made fun of the Irish (pronounced "I-*reesh*"), part-ly for laughs and partly because of stubborn prejudice. A nineteenth-century anti-Catholic political party, the Know Nothings, influenced early Bonner at-titudes that were passed down to my father. But Dad loved performance (and whiskey) and had so much "blarney" and poetry in his personality that I always envisioned him as a traditional village bard in Ireland. Bards perpetuated oral history in songs and poems, a trait that was part of Dad's DNA. He looked the part as well. He ribbed Mom for having Doyle, therefore Irish, in her ancestry and presumed his own ancestry was German and English.

Boy, was he wrong. I have researched our heritage and found some humorous (to the family) material through the DNA feature of Ancestry.com. Dad had much more Irish in his lineage than Mom did. According to my ethnicity inheritance, my father was half Irish, whereas my mother had just a smidgeon of Irish roots. That Doyle name in her lineage was a surname the early Irish gave to dark-haired Danes and other Scandinavians. Mother was a classic dark Norwegian beauty. Bonner, meanwhile, is a fairly common name in Northern Ireland, not a person from Bonn, Germany, as Dad always told us. The ethnic joke's on Dad! I wish he were still around so I could annoy him with this information.

Dad was an enigma when it came to his biases. He judged people by their religion first. Irish, Italian, and Mexican Americans, whom he assumed were mostly Catholics, drew plenty of comments around our home. Some of my best friends in high school were from the nearby Catholic community of Marienthal, and it bothered me deeply when Dad started harping about Catholicism. I used to try to debate him about this prejudice because I couldn't figure out how someone so open-minded in other areas could be so hard baked in this one. Historically though, it was generational in the Bonners' Protestant Northern Irish lineage. Dad's first known North American ancestor, Peter Bonner, set foot in the U.S. in upstate New York in 1815 from a ship that sailed from Belfast, Ireland. Pop's progenitors would have been virulent Irish Protestants carrying a burden of centuries of warfare with Irish Catholics. This historical perspective doesn't excuse the prejudice; it just helps explain it.

But perhaps due to Mom's influence, Pop was not overtly racist toward people based on the color of their skin. Members of Dad's family were, and he learned the institutional racism of our society, but Dad evolved away from racial prejudice. I remember one time Mom confronted Uncle Eldredge about using the N-word, and Dad nodded his head in agreement with her.

His attitudes about race were mainly scientific. As a paleontologist, he knew from fossil evidence that the human species came out of Africa and that all colors of skin, as well as differences in outward appearance like eyes, hair, and noses, were evolutionary adaptations to the environment over time. One time he told me he thought all races would mingle and that in the future all human beings would be "caramel colored." I truly wish he could have met my husband, Carlton, a Black man from Louisiana. I think Dad would have been at first surprised at my choice of a life partner, but then would have accepted him for the human he is—laid-back and intelligent, just like Pop. They would have challenged each other intellectually, and it would have been fun to watch.

End of an Era
1990s–Present

WE USE THE PHRASE "fossil hunter heaven" wryly because Dad didn't believe in the notion of heaven. He thought heaven and its synonyms were an invention of all religions to make people feel better about dying. He was agnostic, simply saying we can't know what happens at death until we get there. Dad abhorred mankind's wars over religion, complaining, "Why fight over where we're going after we're dead? Who knows? Who cares?"

If well-meaning missionaries or "Bible thumpers" knocked on our door, Dad would not ignore them. Far from it. He would engage them in a lively debate. I once overheard this exchange when a visiting Jehovah's Witness church member came to the door in Leoti. Dad opened the door, said hello, listened briefly, then interrupted the visitor with, "Tell me, did Adam have a belly button?" The man answered, "Well, I guess so." Next Dad asked, "Was he made in God's image?" The visitor replied, "Yes, that's what the Bible says." Then Dad said, with finality, "I guess that means God was a placental mammal, then. Good day!" The nonplussed visitor left a brochure and said goodbye.

In the early 1980s, Dad's health started to decline, and he had surgery performed on his heart and circulatory system. In two different Kansas City hospitals, he had a bypass operation, a "carotid artery cleanout," and a Dacron sleeve placed around his abdominal aorta, which was developing an aneurysm. These procedures, particularly the Dacron sleeve, likely extended his life by ten years.

One time, in the spring of 1992, while preparing for a meeting with some friends, Dad took a bath and was too weak to get out of the tub. Realizing he would be stuck there a while, he drained the tub and stayed in it overnight with only a dry towel to keep warm. It wasn't until the afternoon of the next day that his friends came by to see if he was okay. It took him a full month to feel back to normal. He told me all about it on the phone with a very weak voice and couldn't help joking about it. After that, we all insisted he find a way to summon help in an emergency. Since he lived in such an isolated place, the best alternative for this was a Life Alert device that he wore around his neck. Of course, he made fun of the "I've fallen and I can't get up" line in the product's advertising, even though the bathtub experience had been alarming.

A few months later, on the morning of June 3, 1992, he pressed the Life Alert button. This summoned an ambulance that would take him to the nearest major hospital, in Garden City, sixty miles away. It also notified Chuck, the person listed as his emergency contact, who lived twenty-six miles away from Healy, at Keystone. Chuck and Barb arrived at Healy just as Dad was being rolled out on a gurney toward the open ambulance. For a moment, Dad met Chuck's eyes and couldn't help miming a joke of Redd Foxx's from *Sanford and Son,* one of his favorite sitcoms. He dramatically rolled his eyes up in his head as if to say, "Elizabeth, I'm coming to join you, honey!"

I wonder where his thoughts went in the time while he waited for help to arrive. I choose to believe that his last thoughts were of Mom and were couched in a joke. The natural culmination of his life would be reuniting with her, wherever that might be. Chuck and Barb followed the ambulance to Garden City and heard "Code Blue" when they entered the hospital.

The death notice in the Dighton paper read: "Marion Charles Bonner—born May 5, 1911, in Imperial, Nebraska. Passed away June 3, 1992, in Garden City, Kansas, while traveling by ambulance from his home in Healy, Kansas. Buried beside his wife, Margaret Christine Berg Bonner, in the Leoti Cemetery, south of Leoti, Kansas."

After Dad was laid to rest, Chuck commissioned a family friend and well-known limestone sculptor, Pete Felten of Hays, to create a headstone for Mom and Dad out of Fort Hays Limestone, the hardest chalk of the Kansas Cretaceous. Pete Felten is a nationally known limestone sculptor and artist. His Stone Gallery in Hays is open to the public. Felten was commemorated by the Kansas Legislature in 2023, the year of his ninetieth birthday.

Pop would have been happy about the way he exited the world. He told us he did not want a long, drawn-out period in assisted care or hospice. The only ending that would have been better from his point of view would have been to die hunting fossils and, lying flat on his back on the chalk, turn into a perfectly articulated fossil himself (a perspective that concluded the documentary *Just Another Vertebrate*). Dad would have viewed his life as a blip in geological time, or, to use the oft-repeated metaphor he borrowed from paleontologist William King Gregory, his lifespan of eighty-one years would have been an atom in the cigarette paper atop the Empire State Building.

The majority of Dad's life was focused on the thrill of scientific discovery, the possibilities inherent in every step he took along the chalky bluffs. Through his contributions to museums and his influence on his family, Dad lives on in both his fossils and his progeny. Mom's life was also part of the legacy. He had her at his side as his constant helpmate and support.

My parents lie buried in the land they knew best. It's bleak to some, but to plains dwellers, it couldn't be more appropriate. Out here on the flatlands,

Mom and Dad's gravestone of Fort Hays Limestone in the Leoti cemetery, carved by Hays sculptor Pete Felten, shows the marriage of art and science.

there is comfort in being able to see around yourself in three hundred and sixty degrees, where vast sky meets vast land. Contemplating the endless horizon unconsciously makes you breathe deeper. It gives you a feeling of peace. A man of the Great Plains, Dad spent his entire life there: He lived in Chase County, Nebraska, for eight years; Wichita County, Kansas, for fifty-seven years; and Lane County, Kansas, for his last sixteen years on Earth.

The motto at the top of the Bonner gravestone, beside a cloud, reads, "Gone with the Wind," symbolizing a fatalistic view of life that Pop espoused but also a nod to the never-ceasing wind that blows from the Rockies across the High Plains. And it reminds locals of the couple's years spent running the movie theater in Leoti, since that movie was a perennial success for them. On the right side are Dad's dates, 1911–1992, and symbols of his life: a fossil pick and small fossil fish. On Mom's side, on the left, are her dates, 1917–1967, and symbols of her creativity: a spool of thread and a paintbrush. These touches say something personal about each of them in a way that gravestones usually do not. In the period before the headstone, Dad marked Mom's grave with a large angular geode and a septarian concretion about the size of a volleyball. She and Dad had found these large, unusual rocks in two different Kansas locales, and she had kept them in her flower garden in Leoti.

The two of them were defined by the light, wind, and landscapes of western Kansas. Literally, they are now part of the Great Plains deposits. Mom was open and positive, giving of herself to others, principled and faithful. Dad was a driven egotist, a romantic poet of the possible, the bard of the family. He felt compelled to fill up the room whenever he entered.

U.S. Highway 83 runs from the Mexican border at Brownsville, Texas, to the Canadian border with North Dakota. It bisects the continental United States into approximately equal halves and runs the length of the American Great Plains, dividing them in half longitudinally also. When Kansas was under the Western Interior Seaway, the future highway would have roughly bisected that as well. The short-grass High Plains run parallel to the Rocky Mountains, and their eastern boundary was the one-hundredth meridian, "more or less where the tall grasses of the East stopped and the Western short grasses started."[1]

Called the "Main Street of the Great Plains,"[2] Highway 83 divides Scott and Logan counties in Kansas, also cutting through the exposed Cretaceous chalk beds, so it is a well-known throughway for fossil hunters. As it runs through Scott County, it intersects the starting point of the smaller Kansas Highway 4, which turns east to Pop's final haunts in Lane County, the small town of Healy and the tiny town of Shields.

Beginning in 1980, Chuck and Barb began rehabbing an old historic structure on Highway 83 called the Keystone Church, turning the large limestone building into an art gallery, gift shop, and fossil museum. Using recycled materials, they added a second story, a porch, and a greenhouse to the parsonage structure beside the church, and it became their home. Their Keystone Gallery and Fossil Museum is eighteen miles north of Scott City, six miles north of Lake Scott State Park and nine miles west of Monument Rocks, the famous chalk pyramids of western Kansas. The museum is often a jumping-off point for visitors who are interested in viewing these scenic attractions.

Keystone Gallery was opened to the public in 1991. On the back wall of the gallery, Chuck painted a mural that depicts ancient sea creatures that lived in this area of western Kansas 85 million years ago. Chuck has generously allowed me to use his mural to show an artist's interpretation of what Kansas Cretaceous marine animals looked like (see pages 218–219).

Chuck and Barb became friends with author Brad Madsen and artist Ray Troll when they came by Keystone in the summer of 1992 to interview them for their book, *Planet Ocean*. Troll and Madsen's heavily illustrated product is more than a coffee table book. It is a fanciful, clever, metaphorical roll through the countryside that holds the fossils of Kansas.

The author and illustrator missed meeting Dad by one month. In the book, the experience of becoming acquainted with the western Kansas fossil hunters at Keystone and hearing Chuck's reminiscences about his father made an impact:

> Chuck led us into the kitchen of his stone house and showed us pictures of himself, his brother Orville, and his father digging out another good plesiosaur in 1955 when Chuck was about the age of his own son, Logan. "I've got a video of Dad out in the beds," Chuck told us. "I just can't watch it yet. I'll send it to you." His father had been mightily amused by the hubris of humankind and, with the gleam of wisdom in his eyes after eight decades in the Mesozoic, he liked to describe himself as "just another vertebrate." So we drank a toast to Marion Bonner, arguably the last of the old-time fossil hunters, and turned to the business of settling our minds before sleep.[3]

North of the Keystone Gallery and Fossil Museum, located just west of Highway 83 and roughly halfway between Scott City and Oakley, is Little Jerusalem Badlands State Park. This major Cretaceous chalk exposure has been protected so that visitors can walk on hiking trails through the landscape and experience what fossil hunting feels like but cannot take anything home. Roughly halfway between Scott City and Oakley, the park consists of more than two hundred acres of Smoky Hill Chalk badlands on land acquired by the Nature Conservancy. Williston called this site "Castle City," and locals have given it various names over the years, including New Jerusalem for its resemblance to the walled city of Jerusalem. (Dad would have had a field day with the biblical reference to a scientific site.) The Nature Conservancy's website tells potential visitors, "Explore the Kansas you never knew existed."[4]

Describing milestones along Highway 83, in the chapter on U.S. Highway 83 called "The High Plains and the Chalk Beds," the authors of *Roadside Kansas* noted, "Lake Scott is something of an oasis, one of the few large bodies of water in west-central Kansas." Monument Rocks, in Gove County, just east of Highway 83, the authors noted, is "among the most famous of the chalk formations in western Kansas. ... The main cluster of Monument Rocks is part of the Monument Rocks National Natural Landmark."[5]

Monument Rocks National Natural Landmark, located nine miles east down a country road from Keystone Gallery and Fossil Museum, gleams on a bright summer day in 2023. Photo by Carlton Thomas.

Over the years, Chuck and Barb have gained a reputation of their own, both for Kansas tourism and paleontology. Magazine and newspaper journalists have written about Keystone and the fossil hunters who live there. It was included as a site in the Kansas Byways Program, produced by the Kansas Department of Transportation in 2015. For the Byways publication, Dad's old fossil wagon, which Chuck repaired, is featured on the cover, sitting in front of Monument Rocks. One of the more exciting programs to include them was a PBS documentary on *NOVA*, airing in 2015. They were featured in a segment called "Making North America—Life" and interviewed by Smithsonian Natural History Museum curator Kirk Johnson. The program is still viewable on *NOVA*'s website.[6]

Chuck has branched out to form his own network of museum paleontologists. In 2015, he and Barb worked with the Phillip and Patricia Frost Museum of Science in Miami, Florida, to prepare for exhibit another of George Sternberg's fish-within-a-fish specimens. This one was the first one George found, in 1925. They reconstructed the *Gillicus* within the *Xiphactinus*; the vertebrae of the smaller fish are more scattered than in the famous Sternberg Museum specimen but clearly present inside the fossil. The prepared fishes hang today on a wall beside a complete *Pentanogmius* that Chuck also collected for the museum. (Note: The list of museums at the back of this book includes only the museums that have catalogued specimens found, collected, or sent by Dad or my brothers during Dad's lifetime. The list does not contain museums Chuck has worked with since Dad's death.)

It is likely that through his sales to fossil brokers toward the end of his life, some of the fossils Dad collected with Chuck's assistance ended up overseas. However, the Bonner family has no record of actual specimens. In the late 1990s Dad was working on selling crinoids to a German buyer, possibly for a museum. As far as I can determine through documentary evidence, the fossil record for Marion Bonner is contained in eight North American museums. In order of the number of Bonner fossils held in collections and on exhibit, the museums are the Natural History Museum of Los Angeles County (Los Angeles), the Sternberg Museum of Natural History (Hays), the Field Museum of Natural History (Chicago), the KU Biodiversity Institute and Natural History Museum (Lawrence), the Denver Museum of Natural History, the American Museum of Natural History (New York), the Memphis Museum of Science and History, and the Royal Tyrrell Museum (Drumheller, Ontario, Canada).

"The end of an era," is one of those expressions of finality that Dad loved, similar to "gone with the wind." He used it to refer to the horse-and-buggy days, the Plaza Theater years, and the passing of important friends like George Sternberg. The end of an era is an apt phrase for Dad, too. There will not be future paleontologists in the Kansas chalk dealing with landowners, fossils, and museums in quite the same way. The existence of a fossil hunting family in the area is also likely in the past. A big reason for the change is the current emphasis on fossils as commodities and the landowners' concerns about trespassers on their property.

In a geological overview of Kansas for tourists, Dad is included in a summary paragraph about fossil hunters:

> Fossil collecting in the Niobrara in Kansas started in 1868, when the post surgeon at Fort Wallace, in Wallace County, found the remains of a mosasaur. By the early 1870s, fossil-collecting expeditions to the chalk beds were common. Some of the collectors—such as Benjamin Franklin Mudge, O. C. Marsh, and E. D. Cope—were supported by universities and museums, though the competition to claim new fossils spawned bombastic scientific rivalries. Other collectors operated independently. One of these, Marion Bonner, from nearby Leoti, developed a childhood interest

in fossils into a lifelong passion. Though he lacked formal education, Bonner gained renown as an expert on fossils of the Niobrara, and specimens that he collected can be found in museums throughout the country. He and members of his family discovered new species of Cretaceous life that now bear the Bonner name.[7]

In the years that have passed since Dad's death, we Bonner children have become concerned that our father's efforts will be lost to history. To celebrate him locally, Chuck painted an iconic image of him working in the fossil beds. In 2007 the Museum of the Great Plains in Leoti commissioned Chuck to paint murals to show the ancient life of Kansas. One of the scenes is a portrait of Dad working on a *Clidastes* mosasaur, his trenching tools and collection materials in the quarry, and the trusty Chevy Suburban, Spiker, parked on the hill behind him. This painting includes key details: plastic jugs for transporting water to make plaster, an old five-gallon paint can for mixing the plaster, the scrub brush that dots the chalky hill, and even Dad's chalk-covered clothing. The only fantasy touch is the *Pteranodon* flying overhead. But even though that part of the scene is not "actual," it does capture how the human imagination soars when the fossil hunter is out in the chalk beds, plying his trade.

A mural panel painted by Chuck depicts the Old Man unearthing a partial mosasaur. Courtesy of the Museum of the Great Plains, Leoti, Kansas.

In his later life, Pop lived frugally, drawing a modest Social Security check. During his poor years, he bemoaned the rich, claiming life was too short to obsess over money. "The rich get richer and the poor get children," he would say, a spin on Percy Bysshe Shelley's aphorism, "The rich get richer and the poor get poorer."

For a brief time toward the end of his life, selling a few fossils to fossil brokers became part of his subsistence income, but the thought of money was the last thing on his mind when he got bitten by the paleontology bug as a high school student. Even though Dad's era is over, the scholarship around his fossils continues. He would be pleased.

In terms of sheer numbers, there is no way one individual could collect as many specimens as Cope's or Marsh's teams or any large museum staff. If you count only the known catalogued specimens in museums, the number of Dad's fossils is approximately two hundred. However, the fact that the Old Man of the Fossil Beds spent a lifetime searching and specializing in one area of Kansas, making use of his family's many sets of eyes, meant that area was combed well. Even so, the Cretaceous deposits of Kansas are vast and always eroding, so a wealth of discovery is still possible. The beds are not hunted out, as Handel T. Martin's response to Dad's letter of 1929 indicated. The Bonner family has merely scraped the surface.

Dad's sense of self was paramount. When I think about Dad's egotism, I'm reminded of a section from Wallace Stegner's book about his childhood home on the plains. In the chapter called "The Question Mark in the Circle," Stegner scrutinizes the way the plains environment shapes its inhabitants. I was reminded of Dad when I read this passage:

> You don't get out of the wind, but learn to lean and squint against it. You don't escape sky and sun, but wear them in your eyeballs and on your back. You become acutely aware of yourself. The world is very large, the sky even larger, and you are very small. But also the world is flat, empty, nearly abstract, and in its flatness you are a challenging upright thing, as sudden as an exclamation mark, as enigmatic as a question mark.
>
> It is a country to breed mystical people, egocentric people, perhaps poetic people. But not humble ones. At noon the total sun pours on your single head; at sunrise or sunset you throw a shadow a hundred yards long.[8]

The irony of naming this book *Old Man of the Fossil Beds* is that Dad, even up into his eighties, hated the word "old." He would take offense if he heard that word in reference to him or his acquaintances. "*Say!*" he would proclaim, using one of his favorite words of admonishment. "Who're you calling *old*? I'm for the young!" He was drawn to young people his whole life and loved to repeat to them the same stories his children and grandchildren had heard ad infinitum.

To say that he had a dominant personality is a gross understatement. In group settings, Dad ruled the roost. He borrowed liberally from all sources, but if he repeated something often enough, it became his. In one-on-one conversations, he was quite different, often taking a quiet, instructional tone. As a result of being closer to his mother than his father and hailing from a time when gender roles were tightly followed, he was harder, more competitive, and bawdier with his sons and grandsons than with his daughters and granddaughters. His interactions with female family members were usually gentle, funny, and supportive.

Dad and Mom's children were born over a twenty-one-year span, from 1936 through 1957. Where are the Bonner "kids" now? Like cottonwood seeds, we dispersed from the Great Plains to various parts of the country. In 2025, all of us live outside of Wichita County, and only two are still in Kansas. But no matter where we live, our connections to western Kansas and our memories of the fossil beds remain strong. We are becoming old men and women ourselves, grateful for each day we have. All our children are grown, and we are accumulating grandkids.

The eight Bonner children at a family reunion in Fredericksburg, Texas, summer 2004. Back, left to right: Melanie, Steve, Dana, Chuck; front, left to right: Orv, Clare Jane, Chris, Letty. Photo by Angela Buer.

Orville and his wife, Jean, live in Lawrence. Orv retired from his museum position at the University of Kansas in 1997. Their four children, Vincent, Wesley, Boni, and Dixie, live, respectively, in Oakland, California; Lawrence; Overbrook, Kansas; and Ozawkie, Kansas. Orv and Jean have nine grandchildren.

Clare Jane, whose husband, Ray, died in 2005, settled in Briarcliff, Texas. Her four children, Kari, Travis, Amber, and Margaret (Meg), reside in Tucson, Arizona; Spicewood, Texas; Dripping Springs, Texas; and Briarcliff, Texas. Clare Jane has four grandchildren.

Letty and her husband, Tim McGarvin, retired to Green Valley, Arizona. Their four children, Brent, Tammy, Kirby, and Cori, live in Tishomingo, Oklahoma;

Vail, Arizona; Oro Valley, Arizona; and Canyon Lake, Texas. Letty has ten grandchildren and six great-grandchildren.

Steve and his wife, Sonya, were living in Naperville, Illinois, at the time this book was drafted. Sadly, Steve died from an aggressive brain tumor in 2023. Steve and Sonya have two children, Ashley and Seana, who live in Washington, D.C., and Glen Ellyn, Illinois, respectively, and four grandchildren.

Chris and her husband, Terry Weber, reside in Tucson, Arizona; They have two children, Angela and Amy, who live in Mesa, Arizona, and Marina del Rey, California. Chris and Terry have one grandson.

Chuck stayed on the Great Plains, close to the fossil beds. He and his wife, Barbara Shelton, live in rural Logan County, Kansas. Their son, Logan, lives in Seattle, Washington.

Dana and his wife, Ginny (Dougherty), make their home in Tucson, Arizona. Dana has no children; his legacy is art and a large contribution to the Bonner fossil list, especially the Bonner fish-within-a-fish.

My husband, Carlton, and I live in New Orleans. We have three children, Caitlin, Carlton, Jr., and Raekwon, who live, respectively, in Austin, Texas; Pflugerville, Texas; and Baton Rouge, Louisiana. We have four grandsons.

Returning to western Kansas, and in particular the fossil beds, feels primal, even if we're just exploring Monument Rocks. The chalk outcroppings remain a part of us, but the most direct lessons came from the way Mom and Dad raised us. We learned by example that it was okay to be individuals, and that we should also be tolerant of others. Letty gives an apt summary: "What they gave us ought to be the birthright of all children, but some are not so lucky as we were. It was a wonderful way to be raised, a great upbringing. What you realize, as you age, is that we were safe, loved, and educated."

While writing this book, I reconnected with my nieces and nephews, who live all over the country and range in age from their forties to their sixties. My siblings and my older nieces and nephews imparted their memories of Mom. Since I was young when she passed, listening to their stories made me feel closer to her in surprising ways. An unexpected gift also came in reviewing the hundreds of photos she snapped, where I could see her appreciation of life and family through her photographer's eyes.

I was amazed to discover the impact Dad had on all the grandchildren's lives. He knew all seventeen biological grandchildren, and they remember the ritual passing-out of tools in the fossil beds, the songs and poems, the sense of humor. They recall how he would make a big deal of anything they found, even if it was just a shark's tooth or a piece of fin. Whatever the interest of the grandchild, he would listen, encourage, and usually tease them about it. He loved making up silly names for them and carried on Mom's tradition of cooking dinosaur-shaped pancakes when they visited.

All these years since Dad's passing, I still hear him in my mind daily. It is clear to me that western Kansas and my Dad shaped who I am, and this book has yielded surprising personal insights. I did not mention the names of my ex-spouses nor those of my siblings. I did this to respect their privacy (and my own) and also because, in the rearview mirror, our exes were part of our evolution. We have moved on, as have they.

To be completely honest, I have to look at the way my closeness to my father affected my relationships with men. I have been married three times, and looking into that same rearview mirror, I realize why the first two marriages didn't last. First, I didn't know who I was yet. But second, I just wasn't able to give them the respect that I held for my father—the person he was, the parent he was, the depth of feeling, the intellect, the valuing of autonomy over control. That is an almost impossible combination.

But going beyond Dad, to challenge my worldview and move me into new realms of understanding, my husband, Carlton, has many of Pop's traits and more. Carlton's culture and upbringing are significantly different from mine. Our shared experience, though, is growing up poor with strong parents, which makes us contented and grateful for what we have and where we are now. Living with Carlton, I've learned that a solid long-term union starts with attraction, builds to respect, and thrives on a balance of independence and interdependence. That's our formula, perhaps not right for everyone but perfect for us.

For me, respect is a learned value. It didn't come with my environment, where jokes flew endlessly, and making light of everything was encouraged. Along with mockery, in our genetic code perhaps, is the need to question and debate everything. In "maturity," I've learned that in many contexts this kind of behavior is not welcome.

In addition to my evolution as a spouse, my development into a parent and grandparent comes straight from Dad's playbook. Some child psychologists might have called his parenting style benign neglect. I would call it stubborn consistency with a substrata of love. What came naturally to Dad takes a bit more effort for me. I aspire to be a foundation, to convey that I will always be here for our children and grandchildren. To value patience over pushiness, to let people be, let them develop on their own terms.

To my grandsons I will always say: Explore your surroundings—observe and learn. Look inward as much as you need to, but always keep your eyes on the horizon.

Kansas Cretaceous Marine Life

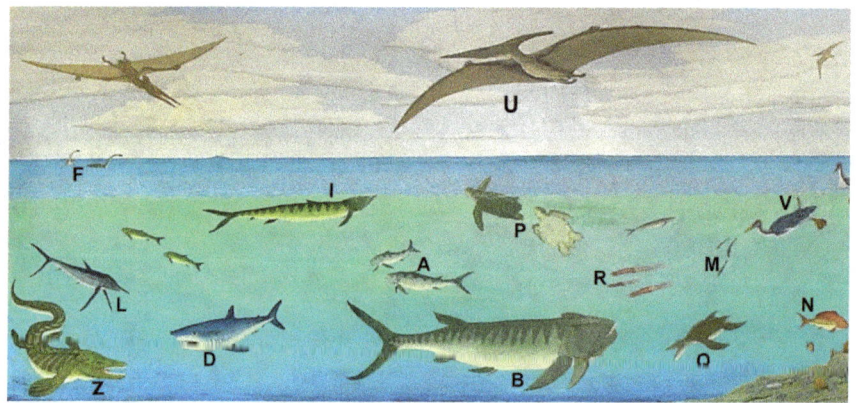

Mural painted by Chuck Bonner is on display at the Keystone Gallery.

KEY (in order of description in the book)

A. *Gillicus*	J. *Ostrea*	S. *Cimolichthys*
B. *Xiphactinus*	K. *Enchodus*	T. *Clidastes*
C. *Squalicorax*	L. *Protosphyraena*	U. *Pteranodon*
D. *Cretoxyrhina*	M. *Apsopelix*	V. *Hesperornis*
E. *Tylosaurus*	N. *Pentanogmius*	W. *Ichthyornis*
F. *Elasmosaurus*	O. *Protostega*	X. *Pachyrhizodus*
G. *Uintacrinus*	P. *Toxochelys*	Y. *Inoceramus*
H. *Ichthyodectes*	Q. *Dolichorhynchops*	Z. *Platecarpus*
I. *Saurodon*	R. *Tusoteuthis*	

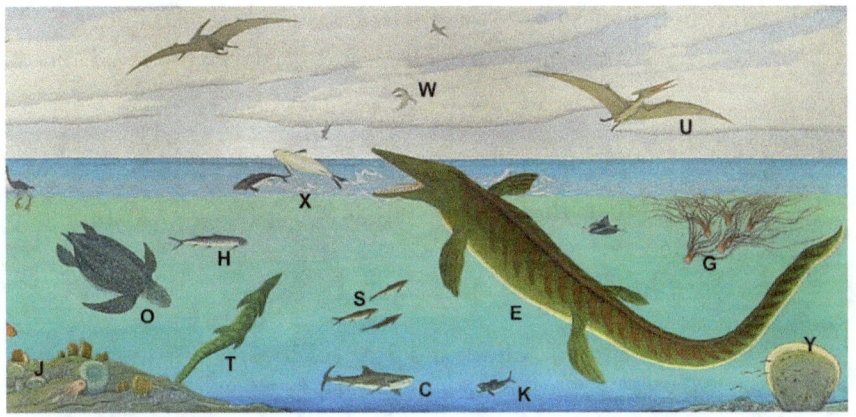

Images are not to scale.

Not shown are junior species, plus: *Archaeolamna, Bonnerichthys, Ctenochelys, Pecten,* and *Polycotylus*

Museums with Bonner Fossils

The list of natural history museums includes only the museums that have catalogued specimens found, collected, or sent by Dad or my brothers during Dad's lifetime. It does not contain museums Chuck has worked with since Dad's death.

- American Museum of Natural History. 200 Central Park West, New York, NY. Holds three fossils collected by Marion Bonner, four by Chuck Bonner.
  www.amnh.org

- Denver Museum of Nature and Science. 2001 Colorado Boulevard, Denver, CO. Holds five fossils collected by Marion Bonner, two by Chuck Bonner.
  www.dmns.org

- Field Museum of Natural History. 1400 S. Lake Shore Drive, Chicago, IL. Holds twenty-three fossils collected by Marion Bonner.
  www.fieldmuseum.org

- Keystone Gallery and Fossil Museum. 401 U.S. Highway 83, Scott City, KS. All fossils at this museum were collected and prepared by Chuck Bonner and Barbara Shelton. One cast created by Chuck Bonner of Marion Bonner's FHSM *Dolichorynchops* paddle is on exhibit.
  www.keystonegallery.com

- KU Biodiversity Institute and Natural History Museum. 1345 Jayhawk Boulevard, Lawrence, KS. Vertebrate Paleontology department holds twelve fossils collected by Marion Bonner, six hundred fifteen collected by Orville Bonner,[1] one by Chuck Bonner.
  www.biodiversity.ku.edu

- Memphis Museum of Science and History. 3050 Central Avenue, Memphis, TN. Holds one major fossil collected by Marion Bonner (*Tylosaurus prorigor*).
  www.moshmemphis.com

- Natural History Museum of Los Angeles County. 900 Exposition Boulevard, Los Angeles, California. Holds ninety-eight fossils collected by Marion Bonner,[2] three by Chuck Bonner, one by Dana Bonner.
  www.nhm.org

- Royal Tyrrell Museum. Midland Provincial Park, Highway 838, four miles northwest of Drumheller, Alberta, Canada. Holds one major fossil collected by Marion Bonner and sons (fish-within-a-fish, *Gillicus* inside a *Xiphactinus*).
  www.tyrrellmuseum.com

- Sternberg Museum of Natural History, 3000 Sternberg Drive, Hays, KS. Holds thirty-one fossils collected by Marion Bonner, twenty-eight collected by Orville Bonner, three by Dana Bonner, and one by Chuck Bonner.
  www.sternberg.fhsu.edu

Acknowledgments

CARLTON, MY PERSONAL BEDROCK, inspired and patiently supported me from the launch of this project through the long months of research, interviews, writing, and editing needed to produce it. He planned soul-restoring vacations to celebrate each chunk of the book's completion, and his technical advice and practical comments kept me grounded. My daughter, Caitlin McColl, provided moral support and enthusiasm, reviewed early drafts, and asked probing questions about her grandparents.

My seven brothers and sisters—Orville, Clare Jane, Letty, Steve, Chris, Chuck, and Dana—happily imparted their memories of Mom, Dad, and the experience of our family growing up in the fossil beds and in a small western Kansas town. I'm especially indebted to Chuck for fossil-specific information and his mural of Cretaceous creatures. And my siblings' support systems, my wonderful sisters- and brothers-in-law, reviewed drafts and offered terrific advice.

In addition, our Berg cousins and my siblings' children, the second generation of Mom and Dad's offspring, added their perspectives and memories. I am grateful to my niece, Ashley Bonner Zung, who read the first draft to my brother Steve before he passed away in September 2023.

Many thanks to Karen Walk, curator of the Museum of the Great Plains in Leoti, Kansas, who shared materials on Bonners, Bergs, and Doyles. The two volumes of *The History of Wichita County, Kansas,* produced by the dedicated members of the Wichita County Historical Society, proved a treasure trove. And Charlie Norton, Dad's Leoti friend who helped him collect *Bonnerichthys*, added some great quotes to the tale. Unfortunately, Charlie died in the summer of 2024 before he could read this book.

I owe deep gratitude to my friend Judy Ryser, developmental editor from Austin, who read two early versions and helped me focus the narrative. Novelist friends from the KU grad school days, Charles Forrest Jones and Sharon Oard Warner, gave me expert writing and publishing advice, and poet and author Denise Low and Kansas State Geologist Emeritus Rex Buchanan reviewed a pre-

press version of the manuscript. Donna Neal, an artist friend from Austin, made global suggestions, advised on photographs, and designed a beautiful cover and imprint. Cyndi Hughes of Booktique Consulting provided crucial publishing advice and proofreading. Over the years, multiple friends, co-workers, neighbors, and extended family members in Texas, Kansas, California, and Louisiana have offered endless support and suggestions.

For paleontological fact-checking and content, I am indebted to J. D. Stewart and Mike Everhart. Now a California-based paleontologist, J. D. is a native Kansan who has known the Bonner family since the 1960s and was instrumental in bringing to light new knowledge through old fossils. Mike is an adjunct curator at the Sternberg Museum whose encyclopedic knowledge of Kansas Cretaceous fossils and his book, *Oceans of Kansas,* supplied essential documentation. Both scientists reviewed the book twice and sent background resources, illustrations, and scholarly papers.

To get here, I wrote two different books. The first was really for my family. It included almost everything everyone remembered about Dad, Mom, and growing up in Leoti, and I rushed to get it drafted before my brother Steve died. Along the way, though, it became clear that non-family readers would be bogged down by a book of such length and detail. I am grateful to reviewers at the University Press of Kansas who suggested I make the book more memoirish, which motivated me to tell a more personal story of my relationship with my father.

Many thanks to these museum personnel who granted access and information about their Bonner holdings:

- Natural History Museum of Los Angeles County: Dr. Xiaoming Wang, curator of vertebrate paleontology; Dr. Luis M. Chiappe, senior vice president, research and collections; Yolanda Bustos, museum archivist and library resources manager; and Dr. Samuel A. McLeod, collections manager of vertebrate paleontology, who sent database lists, accession information, and answered many questions. Dr. Maureen Walsh, collections manager of the Dinosaur Institute, provided images of specimens.

- Sternberg Museum of Natural History, Hays, Kansas: Dr. Aly Baumgartner, former collections manager of paleontology; Dr. Laura E. Wilson, former curator of paleontology; Dr. Reese Barrick, museum director; Greg Walters, exhibits manager; and adjunct curator Michael Everhart.

- Field Museum of Chicago: Dr. William Simpson, head of geological collections and collections manager of fossil vertebrates, emailed a database list of Dad's fossils.

- University of Kansas Biodiversity Institute and Natural History Museum, Lawrence, Kansas: Anne Tangeman, communications coordinator; and Lori Bryn Schlenker, former program coordinator, sent a list of Bonner fossils to my nephew Wesley Bonner at the time of my brother Orville's retirement. In the vertebrate paleontology department, Chris Beard, senior curator, and Hans-Peter Schultze, curator emeritus, provided clarifications.

- American Museum of Natural History, New York: Dr. Alana Gishlick, senior museum specialist, division of paleontology, gave me access to the museum's data.

- Denver Museum of Nature and Science: Dr. David Krause pointed me in the direction of Kristen A. MacKenzie, earth sciences collections manager, who sent a database list of Bonner fossils.

- Museum of Science and History, Memphis, Tennessee: Dr. Rose Basom, former curator of natural science, sent records confirming the acquisition of the complete *Tylosaurus*. Raka Nandi, director of exhibits and collections; and Marilyn Masler, associate registrar, collections department, provided images of this superb fossil.

- Royal Tyrrell Museum, Drumheller, Alberta: Brandon Strilisky, head of collections management, confirmed that Marion Bonner was the collector of the museum's fish-within-a-fish and happily provided museum photos.

Select Bibliography

Aber, James S., Susan E. W. Aber, and Michael J. Everhart. *Roadside Geology of Kansas*. Missoula, MT: Mountain Press, 2023.

Adams, Dawn A. "*Trinacromerum bonneri*, New Species, Last and Fastest Pliosaur of the Western Interior Seaway." *Texas Journal of Science*, 49, 3 (August 1997): 179–198.

Adams, Virginia, et al. *On the Hill: A Photographic History of the University of Kansas*. Lawrence: University Press of Kansas, 2007.

American Association of Petroleum Geologists. "Memorial: Arthur Albert Wedel (1898–1941)." *AAPG Bulletin* (December 1941): 2230–2231.

Bell, Alyssa and Luis M. Chiappe. "The Hesperornithiformes: A Review of the Diversity, Distribution, and Ecology of the Earliest Diving Birds." *Diversity* 14, 267 (April 1, 2022): 1–28.

Bennett, S. Christopher, ed. "Niobrara Chalk Excursion Guidebook: Commemorating the Fiftieth Anniversary of the Society of Vertebrate Paleontology and One Hundred Years of Vertebrate Paleontology at the University of Kansas." Lawrence and Topeka: The Museum of Natural History and the Kansas Geological Survey, 1990.

------. "Taxonomy and Systematics of the Late Cretaceous Pterosaur *Pteranodon* (Pterasauria, Pterodactylia)." *Occasional Papers of the Natural History Museum, University of Kansas* 169 (September 21, 1994): 1–70.

Bessire, Lucas. *Running Out: In Search of Water on the High Plains*. Princeton and Oxford: Princeton University Press, 2021.

Bonner, Orville. "An Osteological Study of *Nyctosaurus* and *Trinacromerum* with a Description of a New Species of *Nyctosaurus*." M.S. thesis, Fort Hays Kansas State College, 1964.

Brown, Lillian. *I Married a Dinosaur*. New York: Dodd, Mead and Company, 1950.

Buchanan, Rex, ed. *Kansas Geology: An Introduction to Landscapes, Rocks, Minerals, and Fossils*. Lawrence: University Press of Kansas, 1984.

Buchanan, Rex C. and James R. McCauley. *Roadside Kansas: A Traveler's Guide to Its Geology and Landmarks*. Lawrence: University Press of Kansas, 1987.

Bush, Larry. "Death Ends His Dedication" (obituary of Claude W. Hibbard). *Ann Arbor (Michigan) News*, October 10, 1973.

Carter, Laura Lee. "Physicians' Promise of a Cure for Tuberculosis Lured Many People West." *Wild West*, December 2008. *https://www.historynet.com/physicians-promise-cure-tuberculosis-lured-many-people-west/*

Colbert, Edwin H. *The Great Dinosaur Hunters and Their Discoveries*. New York: Dover Publications, 1984.

Corn, Mike. "Fossil 'Discovery' Rewrites History." *Hays (Kansas) Daily News*, February 21, 2010.

Cook, Todd D., et al. "A Partial Skeleton of the Late Cretaceous Lamniform Shark, *Archaeolamna kopingensis*, from the Pierre Shale of Western Kansas, U.S.A." *Journal of Vertebrate Paleontology* 31, 1 (February 8, 2011): 8–21.

Cope, E. D. "On the Geology and Paleontology of the Cretaceous Strata of Kansas." Pages 318–349 in *Preliminary Report of the United States Geological Survey of Montana and Portions of Adjacent Territories, Being a Fifth Annual Report of Progress by F. V. Hayden, United States Geologist, Part 3: Paleontology*. Washington, D.C.: U.S. Government Printing Office, 1872.

------. *The Vertebrata of the Cretaceous Formations of the West, Volume 2, F.V. Hayden, ed.* U.S. Geological Survey of Territories. Washington D.C.: U.S. Government Printing Office, 1875.

Davidson, Jane P. "Edward Drinker Cope, Professor Paleozoic, and *Buffalo Land*." *Transactions of the Kansas Academy of Science* 106, 3–4 (2003): 177–191.

Davis, Kenneth S. *Kansas: A Bicentennial History*. New York: W. W. Norton, 1976.

Dighton (Kansas) Herald, June 21, 1989 and June 6, 1992.

Egan, Timothy. *The Worst Hard Time: The Untold Story of Those Who Survived the Great American Dust Bowl*. Boston and New York: Houghton Mifflin, 2006.

Everhart, Michael J. "Larry Dean Martin (1943–2013): Renaissance Paleontologist." *Transactions of the Kansas Academy of Science* 116, 1–2 (2013): 59–62.

------. *Oceans of Kansas: A Natural History of the Western Interior Sea*, 2nd ed. Bloomington: Indiana University Press, 2017.

------. "Oceans of Kansas Paleontology: Fossils from the Late Cretaceous Western Interior Sea." December 20, 2017 (no longer updated). *http://oceansofkansas.com*

------. "William E. Webb: Civil War Correspondent, Railroad Land Baron, Town Founder, Kansas Legislator, Adventurer, Fossil Collector, Author." *Transactions of the Kansas Academy of Science* 119, 2 (2016): 179–192.

Ewing, Mo. "Western Great Plains Shortgrass Prairie: Pawnee Buttes National Grassland, Weld County." Colorado Native Plant Society, 2024. *https://conps.org/project/plains-shortgrass-prairie/*

Fitzgerald, Daniel. *Ghost Towns of Kansas: A Traveler's Guide.* Lawrence: University Press of Kansas, 1988.

Fletcher, Ernest More. *The Wayward Horseman.* Denver, CO: Sage Publishing, 1958.

Friedman, Matt, et al. "100-Million-Year Dynasty of Giant Planktivorous Bony Fishes in the Mesozoic Seas." *Science* 327, February 19, 2010, 990–993.

Friedman, Matt, et al. "Geographic and Stratigraphic Distribution of the Late Cretaceous Suspension-Feeding Bony Fish *Bonnerichthys gladius* (Teleostei, Pachycormiformes)." *Journal of Vertebrate Paleontology* 33, 1 (January 2013): 35–47.

Frazier, Ian. *Great Plains.* New York: Farrar, Straus, Giroux, 1989.

Gardner, James D., Donald M. Henderson, and François Therrien. "Introduction to the Special Issue Commemorating the Thirtieth Anniversary of the Royal Tyrrell Museum of Palaeontology, with a Summary of the Museum's Early History and Its Research Contributions." *Canadian Journal of Earth Sciences* 52, 8 (August 5, 2015): v–xxxiii.

Grande, Lance. *Curators: Behind the Scenes of Natural History Museums.* Chicago: University of Chicago Press, 2017.

Gregg, Josiah. *Commerce of the Prairies,* ed. Max I. Moorhead. Norman: University of Oklahoma Press, 1974.

Gridley, Roy and Karl Gridley, introductory essays. *On the Hill: A Photographic History of the University of Kansas.* Lawrence: University Press of Kansas, 2007.

Gridley, Roy E. "Some Versions of the Primitive and the Pastoral on the Great Plains of America." *Survivals of Pastoral,* ed. Richard F. Hardin. University of Kansas Publications, Humanistic Studies 52, (1979): 61–85.

Hamric, Sharon. "Fossils Feather His Cap: Amateur Digger Makes a Name." *Wichita Eagle-Beacon,* August 11, 1985, 1.

Hanks, Kathy. "Full-Time Fossil Finder," *Kansas!* 2, 1983, 8–9.

Hays (Kansas) Daily News, April 24, 1958, May 26, 1976, February 21, 2010, and June 21, 2010.

Hattin, Donald E. "Stratigraphy and Depositional Environment of Smoky Hill Chalk Member, Niobrara Chalk (Upper Cretaceous) of the Type Area, Western Kansas." *Kansas Geological Survey* 225 (December 1982): 1–108.

Hernandez, Daniela. "Plesiosaurs Carried Young Like a Mammal, Study Finds." *Los Angeles Times*, August 11, 2011.

Hone, David W. E., Mark P. Witton, and Michael B. Habib. "Evidence for the Cretaceous Shark *Cretoxyrhina mantelli* Feeding on the Pterosaur *Pteranodon* from the Niobrara Formation." *PeerJ* 6 (December 14, 2018): e6031. *https://doi.org/10.7717/peerj.6031*

Hunting, Jill. *For Want of Wings: A Bird with Teeth and a Dinosaur in the Family.* Norman: University of Oklahoma Press, 2022.

Hutchinson (Kansas) News, March 5, 2010, August 12, 2011.

Huxley, Thomas. "On a Piece of Chalk." London: *Macmillan's Magazine,* 1868. In The Huxley File, Collected Essays VIII. Clark University, 1998. Accessed September 5, 2024. *http://aleph0.clarku.edu/huxley/CE8/chalk.html*

Jones, Jack. "People and Events: Only in L.A." *Los Angeles Times,* January 20, 1988, 11.

Kansas Jayhawks Athletics, "Traditions: History of the Jayhawk." University of Kansas, 2023. *https://kuathletics.com/sports/2024/4/11/traditions.aspx*

Kansas Geological Survey. "Post World War II." *History of the Kansas Geological Survey.* February 2003. *https://www.kgs.ku.edu/Publications/Bulletins/227/11_post.html*

Kansas Historical Society. *Kansapedia.* "Snow, Francis Huntington," 16875. "Sternberg, George Fryer," 17251. "Wheat," 12235. "Wheat People, Part 7," 10875. Accessed September 5, 2024. *https://www.kshs.org/kansapedia/kansa-pedia/19539*

Kansas State Board of Agriculture. "Niobrara: Fossils." First Biennial Report of the State Board of Agriculture to the Legislature of the State of Kansas, for the Years 1877–1878. "Niobrara—Fossils." Chicago: Rand, McNally Printers and Engravers, 1878.

Kansas Traveling Libraries Commission. "Records of the Kansas Traveling Libraries Commission." Topeka: State Libraries of Kansas, March 19, 2009, 1107–1112.

Kolbert, Elizabeth. *The Sixth Extinction: An Unnatural History.* New York: Henry Holt and Company, 2014.

Konishi, Takuya, et al. *"Platecarpus tympaniticus* (Squamata: Mosasauridae): Osteology of an Exceptionally Preserved Specimen and its Insights into the Acquisition of a Streamlined Body Shape in Mosasaurs." *Journal of Vertebrate Paleontology* 32, 6 (November 2012): 1313–1327.

Lanham, Url. *The Bone Hunters: The Heroic Age of Paleontology in the American West.* New York: Dover Publications. 1973.

Larson, Neal A. "Fossil Coleoids from the Late Cretaceous (Campanian and Maastrichtian) of the Western Interior." *Ferrantia* 59 (2010): 78–113.

Leoti (Kansas) Standard, November 20, 1941, November 27, 1941, March 22, 1988, August 22, 2007.

Lindgren, Johan, et al. "Convergent Evolution in Aquatic Tetrapods: Insights from an Exceptional Fossil Mosasaur." *PLoS ONE* 5, 8 (August 9, 2010): e11998.

Low, Denise, ed. *Confluence: Contemporary Kansas Poetry.* Lawrence, KS: Cottonwood Magazine and Press, 1983.

Low, Denise. *The Turtle's Beating Heart: One Family's Story of Lenape Survival.* Lincoln: University of Nebraska Press, 2017.

Mair, Shirley. "Smile When You Call it Museum." *Macleans,* September 4, 1965, 20.

Marsh, Othniel Charles. *Odontornithes: Monograph of the Extinct Toothed Birds of North America.* Washington: Government Printing Office, 1880.

Martin, Larry D. "The Origins of Birds and of Avian Flight." Chapter in *Current Ornithology* 1. New York: Plenum Press, 1983.

------. "S. W. Williston and the Exploration of the Niobrara Chalk." *Earth Sciences History* 13, 2 (1994) 138–142.

Matsen, Brad and Ray Troll. *Planet Ocean: A Story of Life, the Sea, and Dancing to the Fossil Record.* Berkeley, California: Ten Speed Press, 1994.

May, Stephen J. *Zane Grey: Romancing the West.* Athens, Ohio: Ohio University Press, 1997.

Mechem, Kirke. "The Mythical Jayhawk." *Kansas Historical Quarterly* 13, 1 (February 1944): 1–15.

Miller, Halsey W., Jr. and Myrl V. Walker. "*Enchoteuthis melanae* and *Kansasteuthis lindneri,* New Genera and Species of Teuthids, and a Sepiid from the Niobrara Formation of Kansas." *Transactions of the Kansas Academy of Science* 71, 2 (Summer 1968): 176–183.

Miller, Halsey W., Jr. "Invertebrate Fauna and Environment of Deposition of the Niobrara Formation (Cretaceous) of Kansas." *Fort Hays Studies Science Series* 8 (March 1968), 1–97.

------. "*Niobrarateuthis bonneri,* a New Genus and Species of Squid from the Niobrara Formation of Kansas." *Journal of Paleontology* 31, 4 (July 1957): 809–814.

Miner, Craig. *Kansas: The History of the Sunflower State, 1854–2000.* Lawrence: University Press of Kansas, 2002.

------. *West of Wichita: Settling the High Plains of Kansas, 1865–1890.* Lawrence: University Press of Kansas, 1986.

Morefield, Stephen R. *But the Blood: A Novel, Based on the True Story of America's Bloodiest County Seat Battle.* Leoti, Kansas: Wichita County Historical Society Press, 2022.

"New Plesiosaur Skeleton at Museum One of Only Three on Display in Nation." *Fort Hays State College Leader,* April 24, 1958, 6.

Nicholls, Elizabeth L. and Henry Isaak. "Stratigraphic and Taxonomic Significance of *Tusoteuthis longa* Logan (Coleoidea, Teuthida) from the Pembina Member, Pierre Shale (Campanian), of Manitoba." *Journal of Paleontology* 61, 4 (1987): 727–737.

O'Keefe, F. Robin. "Cranial Anatomy and Taxonomy of *Dolichorhynchops bonneri* New Combination, a Polycotylid (Sauropterygia: Plesiosauria) from the Pierre Shale of Wyoming and South Dakota." *Journal of Vertebrate Paleontology* 28, 3 (July 2008): 664–676.

O'Keefe, F. Robin and Luis M. Chiappe. "Viviparity and K-Selected Life History in a Mesozoic Plesiosaur (Reptilia, Sauropterygia)." *Science* 333, 6044 (August 12, 2011): 870–873.

"*Platecarpus,*" Wikipedia, accessed September 19, 2024. *https://en.wikipedia.or g/wiki/Platecarpus*

Pratt, Linda Ray. *Great Plains Literature.* Lincoln: University of Nebraska Press, 2018.

Public Broadcasting System. "Making North America: Life." *NOVA* (November 11, 2015): season 42, episode 20.

Richmond (Virginia) Times-Dispatch, "Obituaries: Dr. Shelton Pleasants Applegate." August 22, 2005.

Rogers, Katherine. *The Sternberg Fossil Hunters: A Dinosaur Dynasty.* Missoula, MT: Mountain Press Publishing Company, 1999.

Romer, Alfred Sherwood. *Notes and Comments on Vertebrate Paleontology.* Chicago and London: University of Chicago Press, 1968.

Rosin, Elizabeth, Dale Nimz, and Kristen Ottessen. "Now Showing: Historic Theaters and Opera Houses of Kansas," *Kansas Preservation* 27 (March–April 2005): 7–18.

Salina (Kansas) Journal. October 26, 1969, June 1, 1976.

Schulte, Rebecca Ozier. *The Jayhawk: The Story of the University of Kansas's Beloved Mascot.* Lawrence: University Press of Kansas, 2023.

Schultze, Hans-Peter, et al. *Type and Figured Specimens of Fossil Vertebrates in the Collection of the University of Kansas Museum of Natural History.* Lawrence: University of Kansas, Miscellaneous Publications, 1982–1986. Part I Fossil

Fishes, 73 (1982), Part II Fossil Amphibians and Reptiles, 77 (1985), Part III Fossil Birds, 78 (1986), Part IV Fossil Mammals, 79 (1986).

Shankel, Carol, and Barbara Watkins, eds. *Dyche Hall: University of Kansas Natural History Museum 1903-2003*. Lawrence: Historic Mount Oread Fund, Kansas University Endowment Foundation, 2003.

Sharp, Bill and Peggy Sullivan. *The Dashing Kansan: Lewis Lindsay Dyche: The Amazing Adventures of a Nineteenth-Century Naturalist and Explorer*. Kansas City, MO: Harrow Books, 1990.

Shimada, Kenshu and David J. Cicimurri. "Skeletal Anatomy of the Late Cretaceous Shark, *Squalicorax* (Neoselachii: Anacoracidae)." *Paläontologische Zeitschrift* 79, 2 (June 30, 2005): 241–261.

Shimada, Kenshu., Bruce J. Welton, and Douglas J. Long. "A New Fossil Megamouth Shark (Lamniformes, Megachasmidae) from the Oligocene–Miocene of the Western United States." *Journal of Vertebrate Paleontology* 34, 2 (March 2014): 281–290.

Shor, Elizabeth Noble. *Fossils and Flies: The Life of a Compleat Scientist, Samuel Wendell Williston (1851–1918)*. Norman: University of Oklahoma Press, 1971.

Society of Vertebrate Paleontology. "Leadership and Committees," 2024. *https://vertpaleo.org//leadership-committees/*

------. "Romer-Simpson Medal," 2024. *https://vertpaleo.org//romer-simpson-medal/*

------. "SVP Ethics Code," 2024. *https://vertpaleo.org//code-of-conduct/*

"Speaking of Pictures: The Fossil Skeletons of a Fish Within a Fish Give 90-Million-Year-Old Record of a Glutton." *Life*, July 19, 1954, 6.

Stegner, Wallace. *Wolf Willow: A History, a Story, and a Memory of the Last Plains Frontier*. New York: Penguin Books, 2000.

Sternberg, Charles H. *The Life of a Fossil Hunter*. New York: Henry Holt, 1909, reprint Bloomington: Indiana University Press, 1990.

Sternberg, George F. and Myrl V. Walker. "Report on a Plesiosaur Skeleton from Western Kansas." *Transactions of the Kansas Academy of Science* 60, 1 (Spring 1957): 86–87.

Sternberg, George. "Thrills in Fossil Hunting." *The Aerend: A Kansas Quarterly* (1930): 139–153.

Stewart, J. D. "Niobrara Formation Vertebrate Stratigraphy." *Niobrara Chalk Excursion Guidebook*. Lawrence and Topeka: The Museum of Natural History and the Kansas Geological Survey (October 9–10, 1990): 19–30.

------. "Teuthids of the North American Late Cretaceous." Abstract. *Transactions of the Kansas Academy of Science* 79, 3-4 (1976): 94.

The Nature Conservancy. "Places We Protect: Little Jerusalem Badlands State Park, Kansas," 2024. *https://www.nature.org/en-us/get-involved/how-to-help/pl aces-we-protect/little-jerusalem-badlands-state-park/*

Thompson, Lisa. "The National Youth Association (NYA), 1935" (November 18, 2016). *https://livingnewdeal.org/glossary/national-youth-administrati on-nya-1935*

Tollefson, Julie. "Western Vistas Historic Byway." *Byways of Kansas*. Kansas Department of Transportation (2015): 42-44.

Torres, Christopher R., Mark A. Norell, and Julia A. Clarke. "Bird Neurocranial and Body Mass Evolution across the End-Cretaceous Mass Extinction: The Avian Brain Shape Left Other Dinosaurs Behind." *Science Advances* 7, 31 (July 30, 2021): eabg7099.

Tuttle, William M., Jr. *"Daddy's Gone to War": The Second World War in the Lives of America's Children*. New York: Oxford University Press, 1993.

U.S. Bureau of the Census. 2020. Washington, D.C.: Bureau of the Census.

Utley, Robert M. *The Indian Frontier of the American West 1846–1890*. Albuquerque: University of New Mexico Press, 1984.

Warner, Chuck. *Birds, Bones, and Beetles: The Improbable Career and Remarkable Legacy of University of Kansas Naturalist Charles D. Bunker*. Lawrence: University Press of Kansas, 2019.

Webb, W. E. "Air Towns and Their Inhabitants." *Harper's New Monthly Magazine* (November 1875): 828–835.

------. *Buffalo Land: An Authentic Account of the Discoveries, Adventures, and Mishaps of a Scientific and Sporting Party in the Wild West*. Philadelphia: Hubbard Brothers, 1872.

------. "Neb, the Devil's Own." *The Kansas Magazine* 2, 8 (August 1872).

Weiser-Alexander, Kathy. "Genoa, Colorado, and the Wonder Tower." *Legends of America*, June 2023. *https://www.legendsofamerica.com/genoa-colorado-won der-tower/*

Wichita County Historical Association. *History of Wichita County, Kansas*. Newton, KS: Mennonite Press: Two vols: 1980 and 2003.

Wichita (Kansas) Eagle-Beacon, August 11, 1985, February 18, 2010.

Williston, Samuel W. and Alban Stewart. *The University Geological Survey of Kansas. Paleontology, vol 6, part 2: Cretaceous Fishes, Teleosts*. Topeka, KS: W. Y. Morgan, State Printer, 1900.

Williston, Samuel W. *The University Geological Survey of Kansas. Paleontology, vol 4, part 5: Mosasaurs*. Topeka, KS: W. Y. Morgan, State Printer, 1898.

Winchester, Simon. *The Map that Changed the World: William Smith and the Birth of Modern Geology*. New York: Harper Collins, 2001.

"Who Is Dr. Tilly Edinger?" *Timescavengers: Scavenging the Fossil Record for Clues to Earth's Climate and Life.* Accessed August 15, 2024. *https://time scavengers.wpcomstaging.com/tilly-edinger-travel-grant/who-is-tilly-edinger/*

Notes

Plains Vision

1. As of the U.S. Bureau of the Census, 2020, the populations of Leoti, Wichita County, and Kansas were 1,475, 2,152, and 2,940,865, respectively. Chicago's population was 2,746,388. Kansas's name derives from the Kansa, now reorganized as the Kaw Nation. The Wichita people are a confederation of southern plains tribes.

2. This Arthur Capper quote appears in Craig Miner, *Kansas: The History of the Sunflower State, 1854–2000* (Lawrence: University Press of Kansas, 2002), 16.

3. From a song written by Carson Robinson in 1924. Dad most likely heard this number on an old phonograph.

4. W. E. Webb, *Buffalo Land: An Authentic Account of the Discoveries, Adventures, and Mishaps of a Scientific and Sporting Party in the Wild West* (Philadelphia: Hubbard Brothers, 1872).

5. Unattributed quotations in this book are my father's versions, sometimes altered from the original source. I quote his phrases, poems, and songs as I heard them, but where possible cite the likely source. These lyrics are from the traditional Scottish folksong, "Loch Lomond."

6. Linda Ray Pratt, *Great Plains Literature* (Lincoln: University of Nebraska Press, 2018), 6.

Kansas Immigrants

1. Wallace Stegner, *Wolf Willow: A History, a Story, and a Memory of the Last Plains Frontier* (New York: Penguin Books, 2000), 53.

2. Stephen R. Morefield, *But the Blood* (Leoti, KS: Wichita County Historical Society Press, 2022), cover blurb.

3. Craig Miner, *West of Wichita: Settling the High Plains of Kansas, 1865–1890* (Lawrence: University Press of Kansas, 1986), 224.

4. Bernie Kreutzer, "Moving West: Ancient Memories," *History of Wichita County, Kansas,* 2 (Newton, KS: Mennonite Press, 2003), 295. Marion (Skeet) and Jennings would have been teenagers at the time of this work. Volumes 1 and 2 of the *History of Wichita County, Kansas,* published by the Wichita County Historical Society in 1980 and 2003, are the sources for the local references in this book unless otherwise noted. Also, quoted passages from the town newspaper, the *Leoti Standard,* are from these reference works unless otherwise noted.

5. Clyde Blackburn, "A Country Boy Who Came to Town," *History of Wichita County, Kansas,* 1, 294.

Fish Catches Boy

1. The name Leoti High School, a school for the entire county, changed to the Wichita County Community High School (WCCHS) in 1926 and to Wichita County High School (WCHS) in 1966. I use Leoti High School for the early years (my parents' generation) and WCHS for the later years (my siblings' generation) in this book.

2. Simon Winchester, *The Map that Changed the World: William Smith and the Birth of Modern Geology* (New York: Harper Collins Publishers, 2001) 68, 314.

3. Url Lanham, *The Bone Hunters: The Heroic Age of Paleontology in the American West* (New York: Dover Publications, 1973), 12.

4. Michael J. Everhart, *Oceans of Kansas: A Natural History of the Western Interior Sea* (Bloomington: Indiana University Press, 2017), 327–340. Everhart's chapter on dinosaurs explains the known discoveries in Kansas.

5. American Association of Petroleum Geologists, "Memorial: Arthur Albert Wedel," *AAPG Bulletin* (December 1941): 2230-2231.

6. Dr. Kurt Johnson of the Smithsonian Institution cites this perspective in a documentary produced by the Public Broadcasting System, "Making North America: Life," *NOVA* (November 11, 2015): season 42, episode 20.

7. Thomas Huxley, "On a Piece of Chalk," *Macmillan's Magazine,* 1868.

8. The age of the Kansas Cretaceous, determined after radiometric dating in the early twentieth century, was cited as 90 million years ago. Since the 1950s, scientists have revised the age of the Smoky Hill Chalk portion to roughly 87 to 82 million years ago. Also, the "Cretaceous Sea" is now called the "Western Interior Sea."

9. Chalk is composed primarily of the shells (calcium carbonate) of minute, single-celled algae.

10. Joseph Mellick Leidy (1823–1891) was one of the first U.S. vertebrate paleontologists. Before Cope and Marsh became competitors, most of the fossils collected in the Midwest came to Leidy to study. After the Bone Wars began, Leidy turned his attention to parasitology and anatomy. He received his education at the University of Pennsylvania and became a professor of natural history and anatomy.

11. Lanham, *The Bone Hunters*, 264–268.

12. ------, 162.

13. The names of the college at Hays changed over the years: the Western Branch of the Kansas Normal School (1902), the Fort Hays Kansas State Normal School (1914), the Kansas State Teachers College of Hays (1923), Fort Hays Kansas State College (1931), and Fort Hays State University (since 1977). This book uses the current name throughout.

14. Hans-Peter Schultze, et al., *Type and Figured Specimens of Fossil Vertebrates in the Collection of the University of Kansas Museum of Natural History* (Lawrence: University of Kansas, Miscellaneous Publications). Part I Fossil Fishes, 73 (1982), Part II Fossil Amphibians and Reptiles, 77 (1985), Part III Fossil Birds, 78 (1986), Part IV Fossil Mammals, 79 (1986). The number of type specimens is confirmed on the museum's website: https://biodiversity.ku.edu/vertebrate-paleontology/collections

15. See the photo spread of my brother Chuck Bonner's mural on pages 218–219 for interpretations of what the marine animals of this era looked like. I am grateful to Chuck and Michael Everhart for details about each sea creature mentioned in this book, compiled from Keystone Gallery's website and Michael Everhart's *Oceans of Kansas*.

16. E. D. Cope, "On the Geology and Paleontology of the Cretaceous Strata of Kansas," *Preliminary Report of the United States Geological Survey of Montana and Portions of Adjacent Territories, Being a Fifth Annual Report on the Progress by F.V. Hayden, United States Geologist, Part 3, Paleontology* (Washington, D.C.: U.S. Government Printing Office, 1872), 318–349.

17. Samuel W. Williston and Alban Stewart, *The University Geological Survey of Kansas, 6. Paleontology, Part 2, Cretaceous Fishes, Teleosts* (Topeka, KS: W. Y. Morgan, State Printer, 1900).

18. Michael Everhart, contributor, *Handel Tong Martin (1862–1931)* (Lehi, Utah: Ancestry.com / Find a Grave Memorial, 2023), ID: 9049830.

Boy Meets Girl

1. This song was written in 1925 by Jules Cassard, Henry Brunies, and Merrit Brunies, with lyrics by Dudley Mecum.

2. "Beaches" referred to an area where there was a bend in the creek or a riverbed that was dried out and flat, allowing room for dancing.

3. Current dating of these eras is: Precambrian (4,600–2,500 million years ago), Paleozoic (500–280 million years ago), Mesozoic (250–65 million years ago), and Cenozoic (65 million years ago to the present).

4. Timothy Egan, *The Worst Hard Time: The Untold Story of Those Who Survived the Great American Dust Bowl* (Boston, New York: Houghton Mifflin, 2006), 10.

5. Katherine Rogers, *A Dinosaur Dynasty: The Sternberg Fossil Hunters* (Missoula, MT: Mountain Press Publishing Company, 1999), 233.

6. Edwin H. Colbert, *The Great Dinosaur Hunters and Their Discoveries* (New York: Dover Publications, 1984), 185–186.

7. Roy E. Gridley, "Some Versions of the Primitive and the Pastoral on the Great Plains of America," *Survivals of Pastoral,* ed. Richard F. Hardin. University of Kansas Publications, Humanistic Studies 52, (1979): 79. Gridley cites Josiah Gregg, *Commerce of the Prairies,* ed. Max I. Moorhead (Norman: University of Oklahoma Press, 1954).

8. Laura Lee Carter, "Physicians' Promise of a Cure for Tuberculosis Lured Many People West" (*Wild West,* May 19, 2088).

Dust Bowl Survival

1. Ian Frazier, *Great Plains* (New York: Farrar, Straus, Giroux, 1989), 196.

2. Timothy Egan, *The Worst Hard Time* (Boston, New York: Houghton Mifflin, 2006), 9.

3. ------, 9–10.

4. Lisa Thompson, "The National Youth Association (NYA), 1935," (November 18, 2016). https://livingnewdeal.org/glossary/national-youth-administration-nya-1935

5. Karen Walk, "Municipal Auditorium and City Hall WPA Project," *History of Wichita County, Kansas,* 2 (Newton, KS: Mennonite Press, 2003) 34.

6. Frazier, *Great Plains,* 83.

7. Kenneth S. Davis, *Kansas: A Bicentennial History* (New York: W. W. Norton, 1976), 106.

8. A map of "Historic Data" provided by Dad's friend and Leoti historian Clyde Blackburn is printed on the frontispiece of the *History of Wichita County,* 1. It includes three "archaic Indian sites," one "Indian battle site," and a "Fort Wallace scout campsite."

9. "Co-op" is the short name for the farmer's cooperative grain elevators that serve most farming communities. They're particularly important at wheat harvest, since they provide the means of weighing truckloads of wheat that come in and are stored in the elevator until shipment to markets by rail.

10. Michael J. Everhart, *Oceans of Kansas: A Natural History of the Western Interior Sea* (Bloomington: Indiana University Press, 2017), 224–225.

11. Chuck Warner, *Birds, Bones, and Beetles: The Improbable Career and Remarkable Legacy of University of Kansas Naturalist Charles D. Bunker* (Lawrence: University Press of Kansas, 2019), 120.

12. Everhart, *Oceans of Kansas*, 234.

13. ------, 235.

14. ------, 10.

15. The SVP logo was originally created by Margaret Colbert, wife of Dr. Edwin H. Colbert, one of the SVP's early members.

16. Ferdinand Broili (1874–1946) was a German paleontologist. He worked in the Texas Red Beds (Triassic) with Charles H. Sternberg.

17. Letter from George F. Sternberg to "Mr. and Mrs. M. C. Bonner, Leota [*sic*], Kansas." Jim Rouse was one of Sternberg's assistants at the museum, an employee of FHSU. The Sternberg letters quoted throughout this book are copies from the Bonner family archives; we donated the originals to the Sternberg Museum, and they are now archived in the FHSU library.

Fish Tales

1. Dad used this version of this museum's name, as do I in this book. In 1961, the museum was divided into the Los Angeles County Museum of History and Science (LACM) and the Los Angeles County Museum of Art (LACMA). The museum is currently the Natural History Museum of Los Angeles County.

2. "Leoti Man's Fossil Finds on Display in Los Angeles," *Salina Journal,* June 1, 1976, 11.

3. Scott Seirer, "Californians Inspect Western Kansas Fossils," *Hays Daily News,* May 26, 1976, 12.

4. This quote is from my sister Letty, but it reflects a universal experience of children in that era. See: William M. Tuttle, Jr., *Daddy's Gone to War* (New York, Oxford: Oxford University Press, 1993).

5. "SVP Ethics Code," SVP website, vertpaleo.org (McLean, Virginia: Society of Vertebrate Paleontology, 2023).

6. *Albula* was a catch-all term Dad used for small fish measuring one to one-and-a-half feet long. Most of the specimens Dad called *Albula* are now called *Apsopelix*.

7. Ernest Fletcher, *The Wayward Horseman* (Denver, CO: Sage Publishing, 1958), 131.

Old Man of the Army

1. "Jack Dempsey," Hearst Digital Media, October 22, 2021. https://www.biography.com/athletes/jack-dempsey

2. Kansas Geological Survey, "Post World War II," *History of the Kansas Geological Survey,* February 2003. https://www.kgs.ku.edu/Publications/Bulletins/227/11_post.html

3. Michael J. Everhart, *Oceans of Kansas: A Natural History of the Western Interior Sea* (Bloomington: Indiana University Press, 2017), 42.

4. William M. Tuttle, Jr., *Daddy's Gone to War* (New York, Oxford: Oxford University Press, 1993), 29.

5. ------, 46.

6. ------, 154.

7. Kenneth S. Davis, *Kansas: A Bicentennial History* (New York: W. W. Norton, 1976), 199.

8. ------, 200–201.

9. Kathy Weiser-Alexander, "Genoa, Colorado, and the Wonder Tower," *Legends of America,* June 2023. https://www.legendsofamerica.com/genoa-colorado-wonder-tower/

10. Robert M. Utley, *The Indian Frontier of the American West 1846–1890* (Albuquerque: University of New Mexico Press, 1984), 88.

Theater Operations

1. Elizabeth Rosin, Dale Nimz, Kristen Ottessen, "Now Showing: Historic Theaters and Opera Houses of Kansas," *Kansas Preservation* 27 (March–April 2005), 13.

2. Michael J. Everhart, *Oceans of Kansas: A Natural History of the Western Interior Sea* (Bloomington: Indiana University Press, 2017), 134.

3. Michael J. Everhart, *"Pentanogmius (Bananogmius) evolutus,"* Oceans of Kansas Paleontology website, 2007, May 21, 2012. http://oceansofkansas.com/pentanogmius.html

4. Stephen J. May, *Zane Grey: Romancing the West* (Athens, OH: Ohio University Press, 1997), 105.

5. According to the 2020 census, Leoti's 1,475 residents identified as: non-Hispanic white (949), Black or African American (2), Native American (18), Pacific Islander or Native Hawaiian (2), other (253), and two or more races (251). Hispanics or Latinos of any race numbered 585.

Digging the Dirt

1. Kansas Historical Society, "Wheat People, Part 7: Celebrating Kansas Harvest—To Market, To Market." 2024. https://www.kshs.org/p/wheat-people-introduction/10875

2. Rex Buchanan, *Kansas Geology: An Introduction to Landscapes, Rocks, Minerals, and Fossils* (Lawrence: University Press of Kansas / Kansas Geological Survey, 1984), 32. Outcrops around streams in Wichita County are often older than the surrounding Quaternary loess deposits; they are usually designated as Tertiary System (a 60-million-year-old period between the late Cretaceous and the Quaternary). A recent geological overview is J. S. Aber, S. E. W. Aber, and Michael J. Everhart, *Roadside Geology of Kansas* (Sevierville, TN: Mountain Press Publishing, 2023).

3. ------, 35. A more recent analysis of "aquifer depletion in America's heartland" is found in Lucas Bessire's *Running Out: In Search of Water on the High Plains* (Princeton, New Jersey: Princeton University Press, 2021).

4. Dr. Robert Denison (1911–1985) was curator of fossil fishes from 1958 to 1970 at the Field Museum of Natural History in Chicago. He expanded the museum's ichthyology collection from the late 1940s through the mid-1960s and did field work in the U.S., Canada, and Europe.

5. Chimaeroid (ratfish) fossils are exceedingly rare in the Smoky Hill Chalk. They are cartilaginous fishes, but not sharks. Dad sometimes theorized that odd-looking bones such as skates or, later, *Bonnerichthys*, were chimaeroid material.

6. Rex C. Buchanan and James R. McCauley, *Roadside Kansas: A Traveler's Guide to Its Geology and Landmarks* (Lawrence: University Press of Kansas, 1987), 182.

7. Mo Ewing, "Western Great Plains Shortgrass Prairie, Pawnee Buttes National Grassland, Weld County, 2023. https://conps.org/project/plains-shortgrass-prairie/

Grand Man

1. Katherine Rogers, *The Sternberg Fossil Hunters: A Dinosaur Dynasty* (Missoula, MT: Mountain Press Publishing Company, 1999), 163.

2. ------, 176.

3. ------, 163–165.

Rare Friendship

1. Katherine Rogers, *The Sternberg Fossil Hunters: A Dinosaur Dynasty* (Missoula, MT: Mountain Press Publishing Company, 1999), cover note.

2. Joseph Leidy (1823–1891) was an early American paleontologist at the University of Pennsylvania who was a professor of E. D. Cope's. One of Dad's prized possessions, acquired in the 1950s, was a volume published in 1858, titled *Cretaceous Reptiles of the United States,* inscribed by Leidy, the author, to "Dr. F. V. Hayden." Hayden was the lead geologist of the U.S. Geological Survey who wrote and edited *The Vertebrata of the Cretaceous Formations of the West* series in 1875.

3. Michael J. Everhart, *Oceans of Kansas: A Natural History of the Western Interior Sea* (Bloomington: Indiana University Press, 2017), 22.

4. Charles H. Sternberg, *The Life of a Fossil Hunter* (New York: Henry Holt, 1909), 17–18.

5. Kansas State University was called Kansas State Agricultural College at the time; it was founded in Manhattan, Kansas, in 1863 during the Civil War and is the oldest institution of higher learning in Kansas. This book uses the current names, Kansas State University and K-State.

6. George F. Sternberg, "Thrills in Fossil Hunting," *The Aerend: A Kansas Quarterly* (Hays, KS: Kansas State Teachers College, Summer 1930), 146.

7. C. H. Sternberg, *Life of a Fossil Hunter,* 111.

8. Rogers, *Sternberg Fossil Hunters,* 248. Sternberg's three "favorites" were told by Myrl V. Walker to Katherine Rogers in a 1983 interview.

9. "Speaking of Pictures: The Fossil Skeletons of a Fish Within a Fish Give 90-Million-Year-Old Record of a Glutton," *Life,* July 19, 1954, 6.

10. G .F. Sternberg, "Thrills in Fossil Hunting," 141.

11. ------, 148.

12. C. H. Sternberg, *Life of a Fossil Hunter,* 112.

13. ------, 111–112.

14. "New Plesiosaur Skeleton at Museum One of Only Three on Display in Nation," *Fort Hays State College Leader,* April 24, 1958, 6.

15. George F. Sternberg and Myrl V. Walker, "Report on a Plesiosaur Skeleton from Western Kansas" (*Transactions of the Kansas Academy of Science,* vol. 60, no. 1, 1957), 87.

16. Dr. Bobb Shaeffer (1913–2004) was curator of vertebrate paleontology at the American Museum of Natural History, New York. A graduate student under William King Gregory, he was a curator at the AMNH for more than forty years.

17. Everhart, *Oceans of Kansas,* 51.

18. The gladius is a part of a squid located inside the squid's mantle. Since it is primarily composed of chitin (the material insect shells are made of), it was the part of the invertebrate animal that fossilized because minerals replaced the chitin as they did fish scales and bone.

19. Miller, Halsey W., Jr., "*Niobrarateuthis bonneri*, a New Genus and Species of Squid from the Niobrara Formation of Kansas" (*Journal of Paleontology* 31, July 1957): 809–814.

20. Elizabeth L. Nicholls and Henry Isaak, "Stratigraphic and Taxonomic Significance of *Tusoteuthis longa* Logan (Coleoidea, Teuthida) from the Pembina Member, Pierre Shale (Campanian) of Manitoba" (*Journal of Paleontology* 61, 4, 1987): 727–737. Nicholls and Isaak cited earlier works stating the squids were all one species, such as J. D. Stewart, "Teuthids of the North American Late Cretaceous" (*Transactions of the Kansas Academy of Science,* 79, 3–4, (1976), 74.

21. Everhart, *Oceans of Kansas,* 50–51.

22. Lance Grande, *Curators: Behind the Scenes of Natural History Museums* (Chicago: University of Chicago Press, 2017), 383. The 20 percent figure matches current job statistic numbers on career websites.

23. ------, 400.

24. Society of Vertebrate Paleontology, "Leadership and Committees," 2024. https://vertpaleo.org/leadership-committees/

25. Elizabeth Kolbert, *The Sixth Extinction: An Unnatural History* (New York: Picador, 2014), 107–108.

26. Society of Vertebrate Paleontology, "Romer-Simpson Medal," 2024. https://vertpaleo.org/romer-simpson-medal/

27. Michael Everhart told me the museum building was originally a fitness center, foreclosed on for unpaid taxes. The university acquired it, then renovated it into the museum. The large lobby formerly held an Olympic-sized swimming pool.

28. Everhart, *Oceans of Kansas,* 30.

History Lessons

1. "The Ballad of Casey Jones" is a traditional American folk song, circa 1909. "Waiting for a Train" was recorded by Jimmie Rodgers in 1929.

2. The digging line came from "Heigh-Ho," a number in *Snow White and the Seven Dwarfs* in 1937. "I'm Looking Over a Four-Leaf Clover" was written by Mort Dixon and Harry Woods in 1927 and later used in several *Merrie Melodies* shorts.

3. Names, personal information, and parcels owned by specific landowners are withheld. Ownership has changed since Dad was active in the chalk beds.

4. For historical collection information of these types of fishes, see Michael J. Everhart, *Oceans of Kansas: A Natural History of the Western Interior Sea* (Bloomington: Indiana University Press, 2017), 134–136.

5. Dad likely came up with this species name from an old reference book by Cope. The species name for the Smoky Hill Chalk specimens is *Saurodon leanus*. This specimen is the complete *Saurodon* in the Los Angeles Natural History Museum, now called *Saurocephalus lanciformis* (LACM 128118).

6. Everhart, *Oceans of Kansas*, 119.

7. Previously called the Field Museum, it was founded to house objects brought to Chicago for the 1893 World's Columbian Exposition. The major donor to that effort was Chicago businessman Marshall Field, of department store fame. At the time of Dad's correspondence with Denison, the museum was called the Chicago Natural History Museum. In 1967 it returned to its earlier name, Field Museum of Natural History, honoring Marshall Field's nephew, Stanley Field, who ran and helped fund the museum.

8. Kansas Traveling Libraries Commission, "Records of the Kansas Traveling Libraries Commission" (Topeka: State Libraries of Kansas, March 19, 2009), 1107–1112.

9. Jane P. Davidson, "Edward Drinker Cope, Professor Paleozoic, and *Buffalo Land*," *Transactions of the Kansas Academy of Science,* 106 (2003), 3–4.

10. Everhart, *Oceans of Kansas,* 10, 198; also Michael Everhart, "William E. Webb: Civil War Correspondent, Railroad Land Baron, Town Founder, Kansas Legislator, Adventurer, Fossil Collector, Author," *Transactions of the Kansas Academy of Science,* 119, 2 (2016): 179–192.

11. William E. Webb, *Buffalo Land: An Authentic Account of the Discoveries, Adventures, and Mishaps of a Scientific and Sporting Party in the Wild West* (Philadelphia: Hubbard Brothers, 1872). Webb printed passages from Cope's manuscript "On the Geology and Vertebrate Paleontology of the Cretaceous Strata of Kansas" (338–346 and 351–365), which, as Michael Everhart has noted, is an unusual way for Cope's scientific research to be presented.

12. Daniel Fitzgerald, *Ghost Towns of Kansas: A Traveler's Guide* (Lawrence, KS: University Press of Kansas, 1988), 241.

13. William E. Webb, "Air Towns and Their Inhabitants," *Harper's New Monthly Magazine,* November 1875, 830.

14. ------, "Neb, the Devil's Own," *The Kansas Magazine,* 2, 8 (August 1872), 128–133.

15. The Bonner display of western Kansas Native American artifacts hangs in the Keystone Gallery and Fossil Museum, Scott County, Kansas.

16. This information was included in a mimeographed program of the ninth annual meeting of the KAA enclosed with a solicitation letter "To Past Members of the KAA," asking previous members to rejoin the organization.

17. Lillian Brown, *I Married a Dinosaur* (New York: Dodd, Mead & Company, 1950).

18. "Cool Water," by Bob Nolan, was written in 1936. Recorded by the Sons of the Pioneers in 1941.

California Connection

1. Clare Jane's first husband contracted leukemia during his service in the U.S. Army and died in 1957. He did not live to see his daughter born.

2. Virginia Adams, et al., *On the Hill: A Photographic History of the University of Kansas,* Introductory essays by Roy Gridley and Karl Gridley (Lawrence: University Press of Kansas, 2007), 199.

3. Now called CReSIS, for Center for Remote Sensing and Integrated Systems.

4. "Obituaries: Dr. Shelton Pleasants Applegate," *Richmond (Virginia) Times Dispatch,* August 22, 2005.

5. Lance Grande, *Curators: Behind the Scenes of Natural History Museums* (Chicago: University of Chicago Press, 2017), 65–66.

6. Kenshu Shimada, Bruce J. Welton, and Douglas J. Long, "A New Fossil Megamouth Shark (Lamniformes, Megachasmidae) from the Oligocene-Miocene of the Western United States," *Journal of Vertebrate Paleontology* 34, 2 (2014): 281–290.

7. Shirley Mair, "Smile When You Call It Museum," *Macleans,* September 4, 1965, 20.

8. Pierre, South Dakota, was named for Pierre Choteau, Jr., a member of the Saint Louis–based Choteau fur-trading family and a pioneer of steamboat trading on the Missouri River. J. D. Stewart notes that a species of ammonite found in the Niobrara is named after Choteau: *Clioscaphites choteauensis.*

9. Todd D. Cook, et al., "A Partial Skeleton of the Late Cretaceous Lamniform Shark, *Archaeolamna kopingensis,* from the Pierre Shale of Western Kansas," *Journal of Vertebrate Paleontology,* 31, 1 (January 2011): 8–21.

10. "Earth Sciences Division Accession Data for Museum Registrar," *Los Angeles County Museum*, February 4, 1965; I received the list and cover sheet in May 2023 from Dr. Sam McLeod, vertebrate paleontology collections manager at the Natural History Museum of Los Angeles. McLeod and J. D. Stewart, formerly of the museum, confirmed that the museum still has this specimen in storage.

Storms and Sadness

1. Othniel Charles Marsh, *Odontornithes: Monograph of the Extinct Toothed Birds of North America* (Washington, D.C.: Government Printing Office, 1880).

2. Alyssa Bell and Luis M. Chiappe, "The Hesperornithiformes: A Review of the Diversity, Distribution, and Ecology of the Earliest Diving Birds," *Diversity* 14, 267 (April 1, 2022): 267–295.

3. Larry D. Martin, "The Origins of Birds and of Avian Flight," *Current Ornithology* 1 (1983): 105–129. See also Michael J. Everhart, *Oceans of Kansas: A Natural History of the Western Interior Sea* (Bloomington: Indiana University Press, 2017), 309.

4. Jill Hunting, *For Want of Wings: A Bird with Teeth and a Dinosaur in the Family* (Norman: University of Oklahoma Press, 2022).

5. Kansas Jayhawks Athletics, "Traditions: History of the Jayhawk," 2024. https://kuat hletics.com/sports/2024/4/11/traditions.aspx. The year 1899 is confirmed by Rebecca Ozier Schulte in *The Jayhawk: The Story of the University of Kansas's Beloved Mascot* (Lawrence: University Press of Kansas, 2023), 7.

6. This excerpt from the original, "Discovered: Ancestor of *Jayhawkornis Kansasensis*," by Raymond C. Moore, in *Graduate Magazine* 3 (Lawrence, KS: University of Kansas, April 1932), 10, was cited in Kirke Mechem's "The Mythical Jayhawk," *Kansas Historical Quarterly* 13, 1 (February 1944): 1–15.

7. Meade is the county seat of Meade County, a western Kansas county directly south of Lane County on the Oklahoma border.

8. Dad shortened this to "Denver" but the full name of the museum at the time was the Denver Museum of Natural History. It is now the Denver Museum of Nature and Science.

9. This was a song by Matt Monro made popular in 1966 and featured in the movie by the same name about a lion cub named Elsa. I didn't see the movie but had the sheet music.

10. The correct spelling is "Kelaenid." The Kelaenidae are in the class Cephalopoda.

11. Halsey W. Miller, Jr., and Myrl V. Walker, "*Enchoteuthis melanae* and *Kansasteuthis lindneri,* New Genera and Species of Teuthids, and a Sepiid from the Niobrara Formation of Kansas," *Transactions of the Kansas Academy of Science* 71, 2 (Summer 1968): 176–183.

12. J. D. Stewart, "Teuthids of the North American Late Cretaceous," Abstract in *Transactions of the Kansas Academy of Science* 79, 3–4 (1976): 94.

13. Neal A. Larson, "Fossil Coleoids from the Late Cretaceous (Campanian and Maastrichtian) of the Western Interior," *Ferrantia* 59 (2010): 78–113.

14. Elizabeth L. Nicholls and Henry Isaak, "Stratigraphic and Taxonomic Significance of *Tusoteuthis longa* Logan (Coleoidea, Teuthida) from the Pembina Member, Pierre Shale (Campanian), of Manitoba," *Journal of Paleontology* 61, 4 (1987): 727–737.

15. J. D. Stewart informed me that *Enchoteuthis* was maintained as a genus by German scientists in 2024, and it follows that the species name *melanae* may also remain an accepted designation.

16. The museum's current name is the KU Biodiversity Institute and Natural History Museum. Our family called it "Dyche Museum" and "KU Museum," so I use those names in this book.

My Muse

1. Rex C. Buchanan and James R. McCauley, *Roadside Kansas: A Traveler's Guide to Its Geology and Landmarks* (Lawrence: University Press of Kansas, 1987), 181.

2. Carol Shankel and Barbara Watkins, *Dyche Hall: University of Kansas Natural History Museum, 1903–2003* (Lawrence, Kansas: Historic Mount Oread Fund, 2003), 28. The inlaid linoleum floor was designed by Myra Wildish Rising and was commissioned for the reopening of the museum in 1941.

3. Preface by then-director Philip S. Humphrey, in Bill Sharp and Peggy Sullivan, *The Dashing Kansan: Lewis Lindsay Dyche* (Kansas City, MO: Harrow Books, 1990), ix.

4. Roy E. Gridley, "Some Versions of the Primitive and the Pastoral on the Great Plains of America," in *Survivals of Pastoral,* ed. Richard F. Hardin, University of Kansas Publications, *Humanistic Studies* 52 (1979): 61–85.

5. Denise Low, *The Turtle's Beating Heart: One Family's Story of Lenape Survival* (Lincoln: University of Nebraska Press, 2017), 16. Another of her books, *House of Grace, House of Blood: Poems* (Tucson: University of Arizona Press, 2024), provides historic and poetic treatment of the massacre of Christian Delawares (Lenapes) in Ohio.

6. Denise Low, ed. *Confluence: Contemporary Kansas Poetry* (Lawrence, KS: Cottonwood Magazine and Press, 1983), 35.

Institutional Knowledge

1. Orville Bonner, "An Osteological Study of *Nyctosaurus* and *Trinacromerum* with a Description of a New Species of *Nyctosaurus*" (Fort Hays Kansas State College master's thesis, 1964).

2. Larry Bush, "Death Ends His Dedication" (Claude W. Hibbard obituary), *Ann Arbor (Michigan) News,* October 10, 1973.

3. Larry Martin was hired after attaining his doctorate from the University of Kansas in 1972. He was an expert on fossil birds and saber tooth cats. Hans-Peter Schultze, a respected paleoichthyologist, received his doctorate from the University of Tübingen, Germany, in 1965.

4. Carol Shankel and Barbara Watkins, eds., *Dyche Hall: University of Kansas Natural History Museum 1903–2003* (Lawrence: Historic Mount Oread Fund, Kansas University Endowment Foundation, 2003), 7.

5. Chuck Warner, *Birds, Bones, and Beetles: The Improbable Career and Remarkable Legacy of University of Kansas Naturalist Charles D. Bunker* (Lawrence: University Press of Kansas, 2019), 33.

6. Url Lanham, *The Bone Hunters: The Heroic Age of Paleontology in the American West* (New York: Dover Publications, 1973), 90.

7. Warner, *Birds, Bones, and Beetles,* 35. Some of the students and museum workers influenced by Williston were Clarence E. McClung, Charles D. Bunker, Barnum Brown, Elmer Samuel Riggs, and Handel T. Martin. Larry Martin summarized Williston's contributions in "S. W. Williston and the Exploration of the Niobrara Chalk," *Earth Sciences History* 13, 2 (1994): 138–142.

8. Brad Matsen and Ray Troll, *Planet Ocean: A Story of Life, the Sea, and Dancing to the Fossil Record* (Berkeley, CA: Ten Speed Press, 1994), 100.

9. Christopher R. Torres, Mark A. Norell, and Julia A. Clarke, "Bird Neurocranial and Body Mass Evolution across the End-Cretaceous Mass Extinction: The Avian Brain Shape Left Other Dinosaurs Behind," *Science Advances* 7, 31 (July 30, 2021): eabg 7099.

10. Michael J. Everhart, "Larry Dean Martin (1943–2013): Renaissance Paleontologist," *Transactions of the Kansas Academy of Science,* 116, 1–2 (2013): 59–62.

11. Dr. Richard J. Zakrzewski received his master's and doctoral degrees in geology from the University of Michigan. He joined the faculty of Fort Hays State University in 1969 and passed away in 2024.

12. This is Dad's variant (probably from his mother) of an old Scottish folk tune, "Wee Cooper o' Fife." An American variant of the tune is called "Risseldy, Rosseldy," but Dad changed the nonsense syllables.

13. Rex C. Buchanan and James R. McCauley, *Roadside Kansas: A Traveler's Guide to Its Geology and Landmarks* (Lawrence: University Press of Kansas, 1987), 185.

Mystery Fish

1. "Introduction to the Special Issue Commemorating the Thirtieth Anniversary of the Royal Tyrrell Museum of Palaeontology, with a Summary of the Museum's Early History and its Research Contributions," *Canadian Journal of Earth Sciences* 52, 8: v–xxxiii.

2. In addition to Stoss, Dad's interviewer was Kate Moyer and the videographer was Ataullah Zia. Stoss remained friends with Dad for years.

3. Matt Friedman, et al., "100-Million-Year Dynasty of Giant Planktivorous Bony Fishes in the Mesozoic Seas," *Science* 327 (February 19, 2010), 990–993. The group of paleontologists in addition to Friedman were Shimada, Martin, Everhart, Liston, Maltese, and Triebold.

4. Matt Friedman, et al., "Geographic and Stratigraphic Distribution of the Late Cretaceous Suspension-Feeding Bony Fish Bonnerichthys gladius (Teleostei, Pachycormiformes)," *Journal of Vertebrate Paleontology* 33, 1, (January 2013): 35–47.

5. Michael J. Everhart, *Oceans of Kansas: A Natural History of the Western Interior Sea* (Bloomington: Indiana University Press, 2017), 130.

6. ------, 10.

7. Mike Corn, "Fossil 'Discovery' Rewrites History," *Hays (Kansas) Daily News,* February 21, 2010.

The Fossil Record

1. Halsey W. Miller, Jr. "Invertebrate Fauna and Environment of Deposition of the Niobrara (Cretaceous) of Kansas, *Fort Hays Studies, Science Series*, 8 (March 1968): 36.

2. Jack Jones, "Only in L.A.: People and Events," *Los Angeles Times,* January 20, 1988, 11.

3. An important early paper was Takuya Kanishi, et al., "*Platecarpus tympaniticus* (Squamata: Mosasauridae): Osteology of an Exceptionally Preserved Specimen and its Insights into the Acquisition of a Streamlined Body Shape in Mosasaurs," *Journal of Vertebrate Paleontology* 32, 6 (November 2012): 1313–1327.

4. Johan Lindgren, et al., "Convergent Evolution in Aquatic Tetrapods: Insights from an Exceptional Fossil Mosasaur," *PLoS ONE* 5, 8 (August 9, 2010): e11998.

5. "*Platecarpus,*" Wikipedia, accessed September 19, 2024. https://en.wikipedia.org/wiki/Platecarpus

6. Kenshu Shimada and David J. Cicimurri, "Skeletal Anatomy of the Late Cretaceous Shark, *Squalicorax* (Neoselachii: Anacoracidae)," *Paläontologische Zeitschrift* 19, 2 (June 30, 2005): 241–261.

7. Todd D. Cook, et al., "A Partial Skeleton of the Late Cretaceous Lamniform Shark, *Archaeolamna kopingensis*, from the Pierre Shale of Western Kansas, U.S.A." *Journal of Vertebrate Paleontology* 31, 1 (February 8, 2011): 8–21.

8. David W. E. Hone, Mark P. Witton, and Michael B. Habib, "Evidence for the Cretaceous Shark *Cretoxyrhina mantelli* Feeding on the Pterosaur *Pteranodon* from the Niobrara Formation," *PeerJ* 6 (2018): e6031. https://doi.org/10.7717/peerj.6031

9. Michael J. Everhart, *Oceans of Kansas: A Natural History of the Western Interior Sea* (Bloomington: Indiana University Press, 2017), 297, citing S. Christopher Bennett, "Taxonomy and Systematics of the Late Cretaceous Pterosaur *Pteranodon* (Pterasauria, Pterodactylia)," *Occasional Papers of the Natural History Museum, University of Kansas* 169 (September 21, 1994): 1–70.

10. Everhart, *Oceans of Kansas,* 30.

11. D. E. Hattin, "Stratigraphy and Depositional Environment of the Smoky Hill Chalk Member, Niobrara Chalk (Upper Cretaceous) of the Type Area, Western Kansas," *Kansas Geological Survey Bulletin* 225, 1982.

12. Public Broadcasting System, *NOVA,* "Making North America: Life," November 11, 2015.

13. J. D. Stewart, "Niobrara Formation Vertebrate Stratigraphy," *Niobrara Chalk Excursion Guidebook* (Lawrence and Topeka: Museum of Natural History and Kansas Geological Survey, 1990), 19–30.

14. Daniela Hernandez, "Plesiosaurs Carried Young Like a Mammal, Study Finds," *Los Angeles Times,* August 11, 2011.

15. Mike Everhart asserts that my father and Chuck both knew about the baby's bones. He saw photos Chuck took of the specimen that showed completely prepared bones of the fetus.

16. Dawn A. Adams, "*Trinacromerum bonneri,* New Species, Last and Fastest Pliosaur of the Western Interior Seaway," *Texas Journal of Science* 49 (August 1997): 179–198.

17. F. R. O'Keefe, "Cranial Anatomy and Taxonomy of *Dolichorhynchops bonneri* New Combination, a Polycotylid (Sauropterygia: Plesiosauria) from the Pierre Shale of Wyoming and South Dakota," *Journal of Vertebrate Paleontology* 28, 3 (2008): 664–676.

18. Adams, "*Trinacromerum bonneri,*" 191.

End of an Era

1. Ian Frazier, *Great Plains* (New York: Farrar, Straus, Giroux, 1989), 6–7.

2. ------, 211.

3. Brad Matsen and Ray Troll, *Planet Ocean: A Story of Life, the Sea, and Dancing to the Fossil Record* (Berkeley, CA: Ten Speed Press, 1994), 107.

4. The Nature Conservancy, "Places We Protect: Little Jerusalem Badlands State Park, Kansas," 2024. https://www.nature.org/en-us/get-involved/how-to-help/places-we-pr otect/little-jerusalem-badlands-state-park/

5. Rex C. Buchanan and James R. McCauley, *Roadside Kansas: A Traveler's Guide to Its Geology and Landmarks* (Lawrence: University Press of Kansas, 1987),183–185. Jim McCauley was a good friend of my brother Steve's when he was at KU.

6. Public Broadcasting System, "Making North America: Life," *NOVA* (November 11, 2015): season 42, episode 20.

7. Buchanan and McCauley, *Roadside Kansas,* 183.

8. Wallace Stegner, *Wolf Willow: A History, a Story, and a Memory of the Last Plains Frontier* (New York: Penguin Books, 2000), 8.

Museums with Bonner Fossils

1. 1,688 fossils at the University of Kansas museum currently have Orville's name attached as single collector, discoverer, or co-collector.

2. Dr. Sam McLeod clarifies that this is only the number that have been catalogued. The museum still has in storage additional Bonner specimens that have not been prepared or catalogued. This "specimen count" applies to all listed museums.

Before you go...

Thank you so much for reading *Old Man of the Fossil Beds*. I hope you enjoyed this journey into another time and place. I'd greatly appreciate it if you would leave an honest review wherever you bought or downloaded the book. Sharing this book with friends and family is a great way to spread the word about my unique father and his family of fossil hunters. Send any feedback you have to me at *chalklilybooks@gmail.com*.

ABOUT THE AUTHOR

Melanie Bonner Thomas is a western Kansas native and the youngest child of field paleontologist Marion Charles Bonner. She received bachelor's and master's degrees in English from the University of Kansas in Lawrence. While a graduate student, she served as teaching assistant and editor of *Cottonwood Review*, a literary magazine. Her interest in Great Plains literature, history, and science can be traced directly to the influence of her father. This is one of many themes in her memoir/biography, *Old Man of the Fossil Beds*.

After serving as a college English instructor in Kansas and Texas, Thomas spent the bulk of her career as a writer, editor, and communications specialist. She lived in Austin, Texas, from 1986 to 2021. While working there as copy chief and associate editor for *Texas Monthly* magazine and as managing editor of *Hispanic* magazine, she contributed history, archaeology, and paleontology features. She currently lives in New Orleans, Louisiana.

Keep track of upcoming titles from Chalk Lily Books at their website: *www.chalklilybooks.com*

Praise for *Mastering the Art of Letting Go of Emotional Pain & Toxic Relationships*

"*Mastering the Art of Letting Go of Emotional Pain & Toxic Relationships* is both practical and deeply empathetic—a true gem. Dr. Puryear creates a safe space where vulnerability is celebrated as strength, a message especially powerful for teenagers. The hands-on exercises, like writing rituals and boundary-setting, make the lessons tangible and transformative. With its holistic approach that touches on social pressures, relationships, and independence, this book is an invaluable resource for educators, mentors, and anyone seeking healthier emotional balance."

—Marie-Hélène Fasquel, Readers' Favorite

"What I loved most about *Mastering the Art of Letting Go of Emotional Pain & Toxic Relationships* was how it moved beyond theory to provide practical, easy-to-follow exercises I could apply in both personal and professional life. Dr. Puryear writes with honesty and depth, turning each chapter into a step toward healing and self-discovery. The journey felt intimate and transformative, bringing reflection, inspiration, and genuine growth. It's a heartfelt and enlightening guide I highly recommend to teens, young adults, parents, educators, and wellness professionals alike."

—Salina Coria, Readers' Favorite

"*Mastering the Art of Letting Go of Emotional Pain & Toxic Relationships* is a compassionate and transformative guide that helped me rediscover my strength and self-worth. Dr. Puryear beautifully blends practicality with empathy, showing that choosing yourself is an act of deep love. His thoughtful strategies, reflective exercises, and mindfulness practices make healing both accessible and personal. This book reminded me that pain does not define me and that resilience, self-awareness, and emotional intelligence can turn even the hardest moments into profound growth."

—Jessica Barbosa, Readers' Favorite

"*Mastering the Art of Letting Go of Emotional Pain & Toxic Relationships* is a truly empowering and compassionate guide that made me feel both capable and supported. Dr. Puryear's encouraging approach shows that real change is within reach, offering practical tools to release the past and build healthier relationships. His thoughtful guidance fosters trust, self-awareness, and emotional balance, leaving readers with the confidence that peace and freedom are not just possible—they're attainable through mindful, loving choices."

—Asher Syed, Readers' Favorite

"*Mastering the Art of Letting Go of Emotional Pain & Toxic Relationships* is an inspiring and educational guide that makes emotional healing practical and attainable. Dr. Puryear's thoughtful exercises and clear takeaways help readers internalize each lesson and apply it to daily life. By focusing on self-compassion, mindfulness, and empathy, the book offers a grounded approach to reducing stress and anxiety in a fast-paced, demanding world. It's a must-read for anyone feeling weighed down by expectations and ready to reclaim peace of mind."

—Pikasho Deka, Readers' Favorite

"Some of us have been through a revolution just to survive our upbringings, not to mention living in a world filled with corruption, abuse, and lies. It's the resilient who survive, and it's books like *Mastering the Art of Letting Go of Emotional Pain & Toxic Relationships* that give you the tools to not only survive the muck, but prevail and live joyfully no matter what you've been through. Read this book and follow the advice for resilience!"

—Angela Shelton, actress, screenwriter, film director, and producer